THE ILLUSTRATED DICTIONARY OF

MONEY AND MATHEMATICS

Copyright © 1993 Merlion Publishing Ltd
First published 1993 by
Merlion Publishing Ltd
2 Bellinger Close
Greenways Business Park
Chippenham
Wiltshire SN15 1BN
UK

Series editor: Merilyn Holme
Editor: Simone Lefolii

Design: Jane Brett, Steven Hulbert
Illustrations: David Graham; Ken Chatterton, Jeremy Gower and
Matthew White (B.L. Kearley Ltd); Oxford Illustrators Ltd; Jeremy
Pike
Cover illustration: Jeremy Gower (B.L. Kearley Ltd)

Consultant: Laurence Noone, BA, ACA

Printed and bound in Great Britain by BPCC Hazells Ltd

ISBN 1 85737 007 4

THE ILLUSTRATED DICTIONARY OF

MONEY AND
MATHEMATICS

Contributors
Mike Nugent
Jenifer Fellows

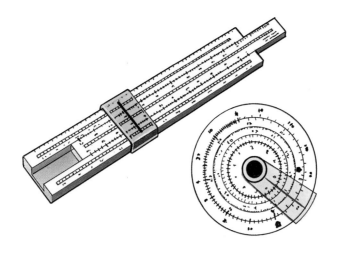

Merlion Publishing

Reader's notes

The entries in this dictionary have several features to help you understand more about the word you are looking up.

- Each entry is introduced by its headword. All the headwords in the dictionary are arranged in alphabetical order.

- Each headword is followed by a part of speech to show whether the word is used as a noun, adjective, verb or prefix.

- Each entry begins with a sentence that uses the headword as its subject.

- Words that are bold in an entry are cross references. You can look them up in this dictionary to find out more information about the topic.

- The sentence in italics at the end of an entry helps you to see how the headword can be used.

- Many of the entries are illustrated. The labels on the illustrations highlight all the key points of information.

- Many of the labels on the illustrations have their own entries in the dictionary and can therefore be used as cross references.

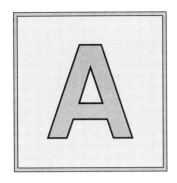

abacus *noun*
An abacus is a device used for counting and doing **arithmetic**. The abacus was invented thousands of years ago and is still in use today. Multiplication is done by repeated addition, and division is done by repeated subtraction.
A skilled person can do arithmetic very quickly on an abacus.

account *noun*
1. An account is a record of financial **transactions**. Businesses keep a set of accounts in a book or on a computer, in which they record money they have paid or are due to pay. An account also records money received or due to be received.
A fixed asset account contains details of fixed assets bought by a business.
2. An account is an arrangement by which a customer can buy goods on **credit** from a supplier. The customer may be an individual or a business.
He has opened an account with Giant Stores.
3. A bank account is an arrangement by which a customer leaves money with a bank and then **withdraws** the money when it is needed.
Money held in a bank account often earns interest for its owner.
4. Accounts are statements recording the **profit** earned by a business, the **assets** owned by the business and the **liabilities** owed by the business. They are usually prepared at regular intervals, perhaps once a year.
The accounts of the company show a profit of £57m earned last year.

accountant *noun*
An accountant is a person who prepares the **accounts** of a business or an individual **client**. The accountant gathers together all the financial information and prepares a summary of the **income** and **expenditure**. An accountant may also provide other services, such as advice on **tax**.
The accountant completed work on the client's accounts for last year.

acre *noun*
An acre is a unit of **area** used to measure land. It is an imperial measure equal to 4,840 square yards and is equal to the metric measure of 4,047 square metres.
The farmer ploughed five acres of land last week.

acute angle *noun*
An acute angle is an angle that is smaller than 90°.
The Leaning Tower of Pisa leans at an acute angle to the ground.

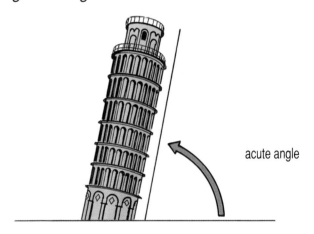

acute angle

addition *noun*
Addition is a kind of arithmetic. It is used to find the total, or **sum**, of two or more numbers put together. The sign for addition is + which is often referred to as 'plus'.
If there are five cars and six trucks, then addition is used to work out that there are eleven vehicles altogether.
At the supermarket we use addition to find out the cost of our shopping.
add *verb*

adjacent angles *plural noun*

Adjacent angles are two angles that meet at the same point, or **vertex**, and which have one side in common.

The adjacent angles measured 26° and 13° each.

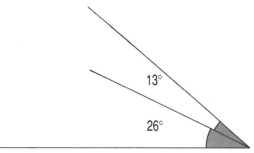

advance *noun*

An advance is a sum of money paid by a **customer** before receiving any goods or services. For example, a business might make advance payments to a building company constructing its new offices. An advance is similar to a **deposit**.

The contract was worth £2m and the building company asked for an advance of £200,000.

afghani *noun*

The afghani is the currency of Afghanistan. An afghani is made up of 100 puls.

algebra *noun*

Algebra is a kind of arithmetic. It is a way of working out number problems using both letters and numbers. The letters can stand for any number. Algebra is used to work out an **equation** such as $y = 2x + 7$.

Algebra was used to work out where the rocket would land.

algebraic *adjective*

algorithm *noun*

An algorithm is a list of all the processes that need to be carried out to solve a problem. Algorithms are used in computer programs because they tell the computer exactly what to do in the correct order.

The scientist used an algorithm to work out the position of the satellite.

alloy *noun*

An alloy is a metal made up of two or more metals which have been mixed together. A combination of copper and zinc makes the alloy brass. Alloys are usually harder, lighter in weight and stronger than the metals from which they are made.

The coins were made from a lightweight alloy.

alternate angles *plural noun*

Alternate angles are the angles made when two lines are crossed by another straight line.

Alternate angles are always in pairs.

altitude *noun*

1. Altitude is another word for **height**. It is usually used to describe how high something is above sea level.

Concorde flies at a top altitude of 16,600 metres.

2. Altitude in geometry is a line drawn downwards from a **vertex** that is **perpendicular** to the opposite side or face of a shape.

A line from the apex of a triangle straight down to the base is an altitude of the triangle.

All three lines measure the altitude of the triangle

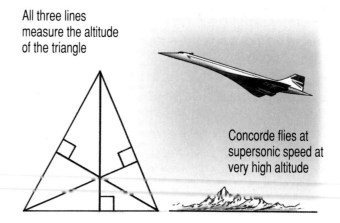

Concorde flies at supersonic speed at very high altitude

AM *abbreviation*

AM or am stands for 'ante meridiem', the Latin words meaning 'before noon'. Six am means six o-clock in the morning.

The timetable showed that the plane left at nine am.

analysis *noun*
In business, analysis means studying a company's **accounts** to decide how healthy it is in financial terms. In most cases, the analysis is used to decide whether it is worth **investing** in a company.
An accountant was asked to do an analysis of the company's accounts.
analyse verb

angle ► page 8

angle of depression *noun*
The angle of depression is the **angle** between a **horizontal** line and a line to a point below the horizontal line.
The angle of depression of the beach from the top of the cliff was 50°.

angle of elevation *noun*
The angle of elevation is the **angle** between a **horizontal** line and a line to a point above the horizontal line.
The angle of elevation from the ground to the top of the tree was 36°.

annual *adjective*
Annual describes something that happens once a year, or which is calculated to last for an entire year. The total of a person's salary for one year is their annual salary.
The tax office looked at the company's annual accounts.

annual percentage rate ► APR

annuity *noun*
1. An annuity is a sum of money which is paid to a person every year.
The company gives their employees an annuity when they retire.
2. An annuity is a kind of **investment**. It earns a yearly income which is paid to the investor.
The annuity paid an income of £1,000 every year for 10 years.

ante-meridiem ► AM

anti-clockwise *adjective*
Anti-clockwise is a movement in the opposite direction to the movement of the hands on a clock face.
To open the door, the key had to turn in an anti-clockwise direction.

apex *noun*
The apex of an object is the highest point it reaches above its **base**. The apex of a tree is its topmost branch on a triangle, or the tip opposite the base.
The apex of a mountain is its highest point, or peak.

applied mathematics *plural noun*
Applied mathematics is **mathematics** that is used to solve practical problems.
Engineers use applied mathematics when they are designing bridges.

approximation *noun*
An approximation is a close guess, but may not be exact. It is close enough to be acceptable.
If there are 495 beads in a box, an approximation of the total would be 500.

APR *abbreviation*
APR is short for Annual Percentage Rate. APR is a way of describing the amount of **interest** that is to be paid on a **loan**.
To decide which loan is cheapest, compare the APRs offered by each bank.

angle *noun*

An angle is the point where two lines or surfaces meet. The angle between two lines measures how much one line turns away from the other. Angles are measured in degrees with a **protractor**. Angles are grouped under particular names depending upon how great the angle is.

At three o'clock, the angle between the hour hand and the minute hand measures 90°.

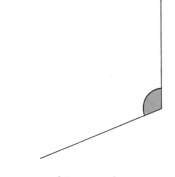

obtuse angle
an angle greater than 90° but less than 180°

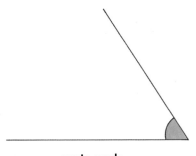

acute angle
an angle greater than 0° but less than 90°

straight angle
an angle of 180°

right angle
an angle of 90°

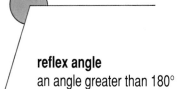

reflex angle
an angle greater than 180° but less than 360°

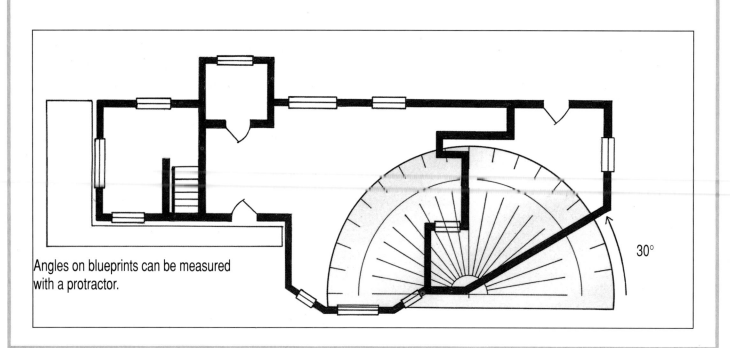

Angles on blueprints can be measured with a protractor.

30°

arc   *noun*
An arc is a section of the **circumference** of a circle, or part of a curve. A pair of **compasses** can be used to draw regular arcs.
A semi-circle is an arc, and so is any portion of the letter S.

area   *noun*
Area is the amount of flat space something takes up on a surface. Area is measured in square units such as square metres or square yards.
The area of the garden is 400 square metres.

arithmetic   *noun*
Arithmetic is a way of working out problems with numbers. Arithmetic uses addition, subtraction, multiplication and division.
Arithmetic was used to work out how much money had been spent.

assay   *noun*
An assay is a test of a metal to see what it contains. It is used to test gold and silver coins or bullion to see how much metal they contain.
The assay showed that the coin contained 2 grams of gold.

asset   *noun*
An asset is an item of value owned by a business or an individual. Money, machines and buildings are all assets. Assets and **liabilities** are listed in the **balance sheet** of a business.
The values of some assets can be hard to determine.

assurance   *noun*
Assurance is the insurance taken out on someone's life. Any individual can take out an assurance policy. A regular payment, or **premium**, is made to an assurance company. If the person dies, the company will pay out a sum of money.
The premium for the assurance policy was £100 per month.

asymmetrical   *adjective*
An asymmetrical object or design is one which has no **symmetry**. There is no way to divide it so that one half is equal in shape and size to the other half.
F, G and J are all letters that have an asymmetrical shape.

auction   *noun*
An auction is a form of public sale where goods are offered for sale without a known price on them. Each item is sold to the person who offers the most money.
The farmer's prize cow was sold for $500 at the auction.

audit   *noun*
An audit is a check on the **accounts** of a company. The audit is carried out by a person from outside the company, usually an accountant. In many countries, companies are required by law to have their accounts audited each year. The person who does the audit is called the auditor.
The audit revealed no major errors in the accounts.

austral   *noun*
The austral is the currency of Argentina. An austral is made up of 100 australes.

Australian dollar   *noun*
The Australian dollar is the currency of Australia. The Australian dollar is made up of 100 cents.

automation   *noun*
Automation is work done by machines with little or no help from people. Computers can be taught how to do many repetitive jobs which people might find boring. Factories are now being built where goods are manufactured automatically, but people are still needed to check that the goods are being made properly.
Robots are often used in automation.

average   *noun*
An average is a number found by adding several values together and then dividing the total by the number of values. The average of 3, 7, 14 and 16 is $3 + 7 + 14 + 16 \div 4 = 10$. Another word for average is mean.
The batsman scored an average of 42 runs.

axis (plural **axes**)   *noun*
1. An axis is a line on a **graph**. The vertical axis, called the y-axis, points upwards. The horizontal axis, called the x-axis, points from left to right.
The vertical axis showed the weight of the bricks.
2. An axis of **symmetry** is a line that divides a shape into two equal, matching parts.
The diameter is an axis of symmetry of a circle.
3. An axis of rotation is a line that a solid body can turn around. The axis of the Earth is an imaginary line from the north pole to the south pole.
The earth turns on its axis once every 24 hours.

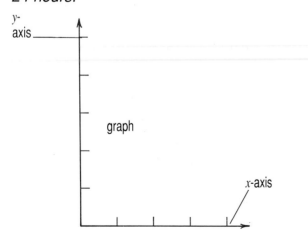

y-axis

graph

x-axis

baht   *noun*
The baht is the currency of Thailand. The baht is made up of 100 satang.

balance   *noun*
1. The balance is an amount of money owed or the amount remaining when the items in an account have been added up.
The balance of the bank account was £250.
2. A balance is a device used to measure the weight of something.
The grocer weighed the vegetables on the balance.

balance of payments   *noun*
The balance of payments is an account totalling all the payments made by a country to other countries, and all the payments received by the country from other countries. A country makes payments to **import** goods from other countries. It receives payments if it **exports** goods. If a country's exports are high and its imports are low, it has a positive balance of payments. If its imports are high and its exports are low, it has a negative, or adverse, balance of payments.
The balance of payments was positive for the third month in a row.

10

balance of trade *noun*
The balance of trade is one part of the **balance of payments**. It refers to money paid and received for **trade** in goods. This kind of trade is known as visible trade. Invisible trade means trade in services, such as banking and insurance.
The balance of trade has improved recently because exports of consumer goods are increasing.

balance sheet ► page 12

bank *noun*
A bank is a place where people can leave their money. These people are the bank's customers and each customer has an **account**. Customers can get their money out again by withdrawing it with a **cheque** or by using a **cash dispenser**.
She paid her money into the bank.

bank note *noun*
A bank note is a printed piece of paper that is issued by a government and which can be used as money. The bank note will be accepted as **legal tender**, which means that everyone who handles it accepts its **face value**. Bank notes often have intricate patterns to prevent **forgery**.
The bank notes in the wallet added up to £25.

banker *noun*
A banker is a person who manages a bank.
The banker loaned £500 to the customer.

bankrupt *adjective*
Bankrupt describes a company or a person who cannot pay their **debts**. When a company goes bankrupt, it will ask a **receiver**, or liquidator, to close down its affairs. Another word for bankrupt is insolvent.
The shoe factory went bankrupt when sales fell badly.

bar *noun*
The bar is the line separating the **numerator** and **denominator** of a **fraction**.
There are three parts to a fraction, the numerator, the bar and the denominator.

bar chart *noun*
A bar chart, or histogram, is a **graph** which uses lines or bars to compare the total value of several items.
The bar chart showed how many people wore size seven shoes.

dozens	1	2	3	4	5.	6
7						
6						
5						
4						
3						

bar code *noun*
A bar code is a pattern of bars and lines which holds information in **code**. A computer is used to read the code, which may contain information on price, quantity and a description of the item.
Many brands of food have a bar code printed on the package.

balance sheet *noun*

A balance sheet is a statement of the financial affairs of a business as at a certain date. It shows the amount of **capital** and retained **profits**, or reserves, invested in the business. It also shows the **assets** which they have been used to buy, and lists the amounts owing by the business.

A balance sheet is often prepared by an accountant.

Balance Sheet of The Beautiful Balloon Company as at 31 March 1992		
ASSETS	£	£
Fixed Assets (at cost less depreciation):		
Land and buildings		500
Plant and machinery		400
Fixtures and fittings		200
Total fixed assets		£1,100
Current assets:		
Stock	600	
Debtors	500	
Prepayments	100	
Cash	200	
	1,400	
Current liabilities:		
Bank loans	200	
Creditors	300	
	500	
Working capital		900
Total net assets		£2,000
CAPITAL AND RESERVES		
Ordinary shares (£1 each)		1,000
Share premium		500
Reserves		500
Total Capital and Reserves		£2,000

things like desks, chairs and telephones

people the company owes money to

subtract the liabilities from the current assets to give the working capital

the totals of each category always balance

12

bargain *noun*

A bargain is something that is bought for a lower price than usual.
A video recorder costing only £25 would be a bargain.

bargain *verb*

To bargain is to discuss what the price of something will be, or how payment will be made.
The customer bargained with the salesperson to try and buy the car cheaply.

barter *verb*

To barter is to trade goods or services without using money. Each person exchanges something they have for something they want.
The farmer bartered three bushels of apples for four chickens

base *noun*

1. The base of a shape is its bottom or lowest line or surface.
The base of the triangle measured eight centimetres.
2. Base is the number of different digits that can be used to write a number. In the decimal system, numbers are made up of the ten digits 0, 1, 2, 3, 4, 5, 6, 7, 8, 9, so the base is 10. In the binary system, every number is made up of the two digits 0 and 1, so the base is 2.
Computers use base 2 arithmetic.

bax unit ► **SI unit**

bazaar *noun*

A bazaar is a kind of **market**. It has shops and stalls where many different kinds of goods are on sale.
Many Middle Eastern towns and cities have bazaars.

bear market *noun*

Bear market is a term which describes the **stock market** when prices are falling. This would suit a 'bear', which is the name given to an investor who sells **shares** in the hope of buying them back at a lower price.
The bear market encouraged some investors to sell their shares quickly.

bearing *noun*

A bearing is a direction measured in **degrees**. It is always the measurement of the angle between a line running north to south and a second line. The bearing of a ship is the angle between the direction that the ship is sailing in and the direction of the north or south pole.
The ship's bearing is 50° and the plane's bearing is 280°.

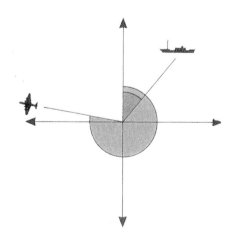

Belgian franc *noun*

The Belgian franc is the currency of Belgium. One Belgian franc is made up of 100 centimes.

bet *noun*
A bet is a kind of agreement between two people. A bet is made on the outcome of something uncertain. For example, if a coin is tossed, one person might bet that the coin will show heads, and the other person will bet that it will show tails. The person who guesses correctly wins the bet.
People sometimes place a bet for money, so that the loser has to pay the winner.
bet *verb*

bid *noun*
A bid is an offer of a sum of money for something. At an **auction**, anyone wishing to buy an item places a bid, according to the amount they are willing to pay. The item will be sold to the person who offers the most money.
The art collector bid £4,000 for the picture.

bill *noun*
1. A bill is a list of prices charged for work done. For example, a motor mechanic may prepare a bill listing the repair work done on a car and the parts replaced. The total of the charges must then be paid by the owner of the car. Another word for a bill is an invoice.
The customer was surprised when he saw the amount of the bill.
2. A bill is a document which promises to pay someone a certain amount of money on a particular date. This is often referred to as a bill of exchange.
The bill of exchange was payable 90 days after it was issued.

bill of lading *noun*
A bill of lading is a list of goods. It describes all the things that are to be transported from one place to another during a single journey.
The ship's captain gave the bill of lading to the customs officer.

billion *noun*
A billion is a thousand million in most countries. It is written 1,000,000,000. In some countries in Europe, a billion is used to mean a million million (1,000,000,000,000).
The company earned profits of two billion pounds last year.

billionaire *noun*
A billionaire is a person who has a billion units of currency. In the United States of America, a billionaire has a billion dollars.
The British billionaire had a billion pounds.

binary *adjective*
Binary describes counting in twos and by **powers** of two. Binary counting uses only the two **digits** 0 and 1. In binary counting, 11011 means 1 sixteen, 1 eight, 0 fours, 1 two and 1 unit. 11011 in binary is equivalent to $16 + 8 + 0 + 2 + 1 = 27$ in decimal counting.
The binary system of counting is used in digital computers.

bisect *verb*
To bisect is to divide into two, usually into equal parts. Lines and angles are bisected when another line is drawn to divide them in half.
The diameter bisects a circle.

bit *noun*
Bit is short for **binary** digit. A bit is a 0 or a 1 that is used in binary numbers.
The number 110101 has 6 bits.

1 1 0 1 0 1

black market   *noun*

A black market is set up so that goods may be sold illegally. Some goods are rationed, or only sold in small quantities, when they are in short supply. Other goods are not allowed to be sold because they are dangerous or rare. However, people may still be able to buy these things on the black market.
Extra sugar was for sale on the black market.

block graph   *noun*

A block graph is a **graph** which uses rectangles, or blocks, to compare the value of different items.
The teacher drew a block graph of the exam results for the class.

board of directors ▶ page 16

boiling point   *noun*

The boiling point is the **temperature** at which a liquid boils. When the temperature is higher than the boiling point, the liquid turns into a gas.
The boiling point of water is 100 degrees Celsius.

bolívar   *noun*

The bolívar is the currency of Venezuela. One bolívar is made up of 100 centimes.

Bolivian peso   *noun*

The Bolivian peso is the currency of Bolivia. One Bolivian peso is made up of 100 centavos.

bond   *noun*

A bond is a printed piece of paper usually issued by a government. Individuals or businesses can pay money to the government to buy a bond. At some time in the future the government will repay the money, and in the meantime it pays **interest** as a reward for lending the money.
The £20,000 worth of bonds earned interest throughout the year.

bonus   *noun*

A bonus is extra money paid as a reward. In some companies, people are paid a bonus if they work harder.
Everyone was paid a bonus for selling more toys than in the previous month.

bookkeeping   *noun*

Bookkeeping is a system of recording the **transactions** of a business by making entries in **books of account**. It is usual to use two entries to describe any transaction, and the term double entry bookkeeping is often used. Today, bookkeeping is often done on a computer, with no real books being used.
The company's bookkeeping system was fully computerized.

Date	Received From	Amount £	Date	Paid To	Amount £
Dec 12	Mr Brown	98.50	Dec 1	dairy	6.00
			Dec 1	Mr White	12.00
Dec 16	Mr Green	1,300.00	Dec 2	petty cash	5.00
			Dec 10	salary	65.00
			Dec 11	post office	4.60
Dec 21	Mrs Black	3,000.00	Dec 17	salary	65.00
			Dec 22	Paper Ltd	125.00
				Film Ltd	600.00
				telephone	300.00
				salary	130.00

board of directors   *noun*

A board of directors is a group of people who manage the business of a **company**. They hold regular meetings, called board meetings, to discuss **policy** and make decisions about what the company should be doing. Each member of the board usually represents one particular branch, or department of the business, such as sales, personnel or distribution.

The board of directors of Superco Inc. meet every month.

Marketing Director responsible for letting the customers know about the product

Personnel Director responsible for the people who work for the company

Executive Director head of the company but not involved in day to day management

Secretary takes notes and creates a record called minutes, of all meetings

Distribution Director responsible for the distribution to customers of the goods the company manufactures

Managing Director head of the company and responsible for its day to day management

Sales Director responsible for selling to the customers the product the company makes

Art Director responsible for the design and production of packaging and any sales or marketing brochures

Production Director responsible for the actual making, or production, of the product in a factory or office

Finance Director responsible for managing the company's money and obtaining extra finance when needed

books of account   *noun*

Books of account are all the records kept by a company to describe its financial activities. For example, sales made to customers may be recorded in a **cash book**. This is one of the company's books of account.
Books of account are written up regularly so that all the transactions of a company are recorded.

borrow   *verb*

To borrow means to obtain money which has to be repaid in the future. People often borrow money from banks. Until the money is repaid, the person who borrows usually has to pay **interest**.
The sailor will borrow £20,000 to buy a new boat.

borrower   *noun*

A borrower is someone who borrows money.
The borrower paid interest each month until the money was repaid.

brackets   *plural noun*

Brackets are symbols used in mathematics to group numbers or letters together. In **algebra**, brackets are used to show which part of an **equation** needs to be worked first. Curly brackets are used to contain the members of a **set**.
Brackets can be curved ( ), square [ ] or curly { }.

brand   *noun*

A brand is a kind of **trade mark**. It is the name that a manufacturing company gives to one of its products.
The shop sold more than one brand of soap powder.
brand   *verb*

breadth   *noun*

Breadth is another word for width. It is the measurement of **distance** from one side of something to the other.
He measured the breadth of the football pitch.

broker   *noun*

A broker is a person who buys and sells things on behalf of other people. An insurance broker will arrange an insurance policy for a customer. A broker makes money by charging **commission** for services.
A stockbroker will buy and sell shares at a stock exchange.

budget   *noun*

1. A budget is an estimate of amounts of money expected to be paid and received during some future period. For example, a business will often prepare a budget to set out its financial plans for the year ahead. In many countries, the government prepares a budget each year. This shows the government's plans for raising money from **taxes**, and spending money on things such as roads and hospitals.
The company's budget for next year shows an expected profit of £80,000.
2. A budget is an amount of money set aside for a particular purpose in the future.
A budget of two million pounds was set for the building project.
budgetary   *adjective*

building society   *noun*

A building society is similar to a bank. Its main activity is to lend money to individuals so that they can buy somewhere to live. Building societies also offer other services like those of a bank. For example, most building societies issue **cheque** books and **cash cards**.
They borrowed £50,000 from a building society to buy a new house.

bull market   *noun*

Bull market is a term which describes the stock market when prices are rising. This would suit 'bulls', or investors who buy **shares** in the hope that they can sell again at a higher price.
The investor held on to the shares as the bull market continued.

bullion *noun*
Bullion is gold or silver before it is made into coins. Bars of gold or silver are bullion.
The United States of America keeps its gold bullion at Fort Knox.

bureau de change *noun*
A bureau de change is a place where the **currency** of one country can be exchanged for the currency of another country.
The traveller converted 50 francs to dollars at the bureau de change.

burglary *noun*
A burglary is the theft of possessions from a building. A burglary can take place from someone's house, or from an office or a factory. Burglary is a crime.
The family lost their television in the burglary.

business *noun*
1. Business is the work that people do. For example, the business of a shopkeeper is to sell things to customers.
The company's business is to make computers.
2. A business is an **organization** set up to supply goods or services to customers in exchange for money. A business may be very small, such as a trader selling clothes from a market stall. It may be very large, like an international bank. Most large businesses, and some small ones, are **companies**.
They set up a new business for publishing books.

buy *verb*
To buy is to give a sum of money for something. People usually buy things at a store.
The carpenter wanted to buy a new saw.

buyer *noun*
A buyer is someone who buys things on behalf of a business. For example, a clothes manufacturer might employ a buyer to buy the material needed to make the clothes.
The buyer bought enough material for 800 coats.

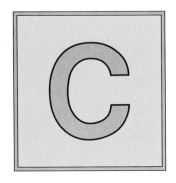

calculation *noun*

A calculation is a way of solving a number problem. Calculations can use addition, subtraction, multiplication and division to produce an answer.
The calculation of the profit took a long time.
calculate *verb*

calculator *noun*

A calculator is a machine that can do **arithmetic**. For a long time the **abacus** and **slide rule** were used to solve problems in arithmetic, but in many countries these have been replaced by the electronic calculator.
Computers are the most powerful calculators.

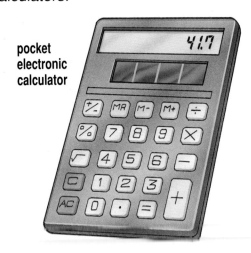

pocket electronic calculator

calculus *noun*

Calculus is a way of calculating in mathematics. It often involves the use of symbols from **algebra** to solve problems concerned with **statistics** and changing quantities, or amounts.
The mathematician used calculus to solve a problem.

calendar *noun*

A calendar is a table that shows the days, weeks and months of a year. Many calendars have spaces next to each date for recording important information, such as appointments and birthdays.
The Hindu calendar is circular and based on lunar months.

Canadian dollar *noun*

The Canadian dollar is the currency of Canada. It is made up of 100 cents.

cancel *verb*

To cancel is to simplify a fraction down to its lowest form. In cancelling, both the numerator and denominator are divided by the same number.
She used cancelling to simplify $\frac{4}{20}$ down to $\frac{1}{5}$.

capacity *noun*

Capacity is the amount of space or **volume** that something which can be poured takes up. The main metric unit of capacity is the litre.
The capacity of the watering can was eight litres.

capital *noun*

The capital of a business is the amount of money invested in the business. This money is spent to buy the things the business needs to carry on its trade. For example, it may be used to buy equipment and buildings.
The capital of the business increased because the business earned higher profits.

capital gain   *noun*
A capital gain is a **profit** earned by selling a major **asset** for more than it cost to buy. Someone might make a capital gain by selling a valuable item, such as a painting. A company might make a capital gain by selling a **fixed asset**. The opposite of a capital gain is a capital loss.
A capital gain of £2,000,000 was made after a racehorse bought for £1,000,000 was sold for double the sum.

capital loss ► **capital gain**

capitalism   *noun*
Capitalism is a system of **economics** where capital, or money, can be used to develop new industry or services. In this system, people or companies can own land, factories or industries to produce goods or services for **profit**. The profit can then be re-invested or distributed to **employees** or **shareholders**. Capitalism allows companies to compete with each other for trade.
Capitalism is considered a good system by people who enjoy trading freely.
capitalist   *adjective*

cartel   *noun*
A cartel is a group of companies which agree to work together so that between them they can dominate a **market**. Often the aim of a cartel is to drive other companies out of the market and establish a **monopoly**. In many countries cartels are illegal.
Three transport companies formed a cartel so as to increase the prices charged to their customers.

cash   *noun*
Cash is money in the form of notes and coins. People sometimes pay for expensive purchases with a **cheque** or with a **credit card**, rather than carry a large amount of cash around with them. Money held in a bank account is also referred to as cash.
They had no cash so they had to pay by cheque.

cash book   *noun*
A cash book is a record, or **account**, which lists all the sums of cash a business receives or pays. This includes payments by cheque as well as by notes and coins.
The cash book showed a balance of $100 at the end of the day.

cash card   *noun*
A cash card is a small plastic card which is used to draw money from a **cash dispenser**. The card contains information about the owner's bank account, so the cash dispenser can tell which account the money is to be taken from.
A cash card is necessary to withdraw money from a cash dispenser.

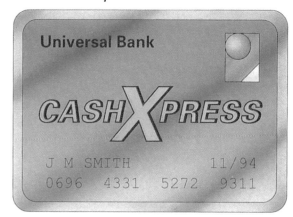

cash dispenser ► page 22

cash flow   *noun*
Cash flow is the amount of money that comes into and goes out of a business.
The company had problems with its cash flow because it had not received the money it was owed.

cash register   *noun*
A cash register is a machine used by a shop to record and hold the money people pay for goods. It has a keyboard which is used to record the cash received from customers, and a drawer to keep it in. A cash register is often connected to a computer which can total the day's takings.
The cash register showed that the customer paid cash for the tie.

cash dispenser *noun*

A cash dispenser is a machine normally operated by a bank or building society, from which customers can take money. The machine is connected to a computer. When a customer puts in a cash card, the computer can tell which **account** the money is to be taken from. The customer must type in the correct PIN, or personal identity number, before the computer will release the cash. The computer notes how much money is taken out.

He used the cash dispenser to withdraw $50 from his account.

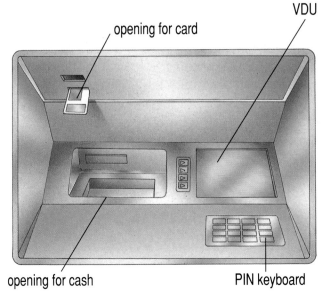

VDU

opening for card

opening for cash

PIN keyboard

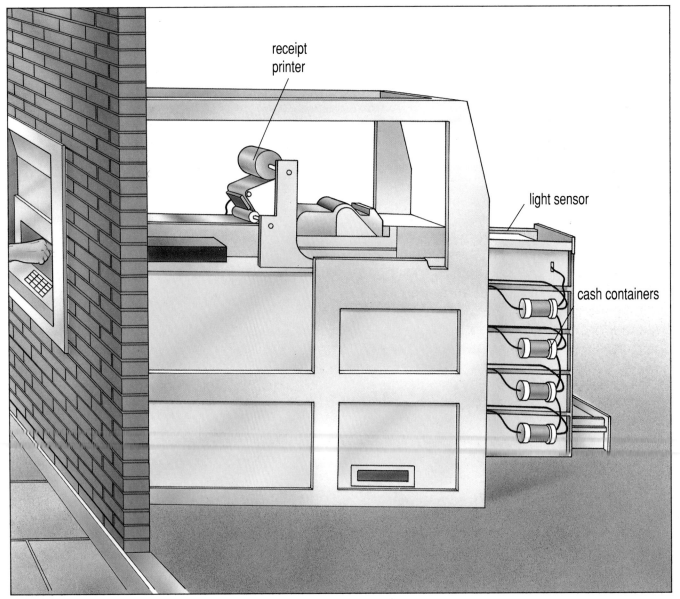

receipt printer

light sensor

cash containers

casino *noun*
A casino is a building where people play games to try to win money. This is known as gambling. People who play card games bet on how their hand of cards will turn out. Roulette is a game in which a small ball spins round a numbered wheel. Players bet on which number the ball will land on when the wheel stops spinning.
The casino was built in the centre of the holiday resort.

Celsius *noun*
Celsius is a scale of temperature in the metric system. 0° Celsius is the freezing point of water and 100° Celsius is the boiling point of water. The Celsius scale is sometimes called the centigrade scale.
The Celsius scale is named after a Swedish scientist.

cent *noun*
A cent is one hundredth of a dollar.

centi- *prefix*
Centi- is a prefix that means one hundredth. A centimetre is a hundredth of a metre, so there are 100 centimetres in a metre.
The cactus was 14 centimetres tall.

centigrade ▶ **Celsius**

centimetre *noun*
A centimetre is a measure of length in the metric system. There are 100 centimetres in a metre. The short form for centimetre is cm.
He measured the table top in centimetres.

century *noun*
A century means 100 years. The name 'century' comes from the Latin word 'centura', which means a hundred.
The church was exactly a century old.

Chamber of Commerce *noun*
A Chamber of Commerce is an organization that helps business people to meet and work together in their local area. It also tries to attract outside clients to work with local suppliers.
The Chamber of Commerce helped them to meet new clients in their town.

chance *noun*
Chance is the way some things seem to happen without a reason. The word chance is also sometimes used to mean **probability**.
By chance, they were wearing the same coloured dress.

change *verb*
To change is to convert from one unit of measure to another. For example, imperial measurements can be changed to metric measurements using **conversion** tables.
They changed their dollars into pesos before they went on holiday.

change *noun*
1. Change is another word for loose coins.
The vending machine only accepts change.
2. Change describes the cash difference between money paid for a purchase and its price. Change is handed back to the customer by the seller.
She received a dollar in change from the shopkeeper.

charge *noun*
A charge is the price asked for in return for providing goods or services.
A charge of $20 was made for the job.

charge *verb*
1. To charge is to ask for payment in return for providing goods or services.
The plumber will charge £45 for the work.
2. To charge is to offer payment by means of a **credit card**.
She was asked if she wanted to pay cash for the item or wanted to charge it to her card.

chart *noun*
1. A chart is a table, diagram or graph used to present information in a visual way. Charts can be used to illustrate information, such as quantity or value.
The chart shows how many sales the company had made in a year.
2. A chart is a map used for navigating in the air or on the water.
The sailor used his chart to find out where dangerous rocks were located.

cheap *adjective*
Cheap describes something which costs less than you might expect.
The computer was cheap at £200.

cheque *noun*
A cheque is a printed piece of paper which orders a bank to pay out money from an **account** held by the owner of the cheque. The cheque has the details of the owner's bank account printed on it. An individual or a business can use a cheque either to **withdraw** cash from the bank in the form of notes and coins, or to pay other people for goods or services supplied.
The owner of the business paid the employees by cheque.

Chilean peso *noun*
The Chilean peso is the currency of Chile. The Chilean peso is made up of 100 centavos.

chord *noun*
A chord is a straight line that joins two points on the **circumference** of a circle. It does not pass through the centre of the circle.
The chord cut the circle into two equal parts.

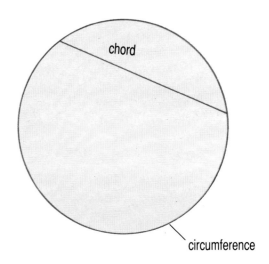

cipher *noun*
A cipher is a secret **code**. Banks often use a cipher to send information to other banks. The cipher is used so that if the message reaches the wrong person by accident, it cannot be understood.
The bank manager used a cipher so the message could only be understood by the person to whom it was sent.

circle *noun*
A circle is a round **shape**. The edge, or **circumference**, of a circle is always the same distance from the centre point.
She drew a circle to show the shape of a wheel.
circular *adjective*

circulation *noun*
Circulation is the way money passes from one person to another. For example, when people buy goods, they give money to a shopkeeper. The shopkeeper then uses this money to buy other goods to sell and also to pay staff wages. In this way, money is passed from one person to another.
The amount of money in circulation changes from year to year.

circumference   *noun*
The circumference of a circle is the distance measured around its edge. If the circumference of a circle is divided by the **diameter**, the answer is always the same. This number is called pi and is written π.
The circumference of the circle is 21 centimetres.

client   *noun*
A client is another word for customer. Clients pay someone else to do a job for them.
The accountant prepared the annual accounts for each client.

clinometer   *noun*
A clinometer is a device used to measure a slope or incline. A clinometer can also be used to measure the **angle of elevation** when working out the height of something.
A clinometer can be used to calculate the height of the Eiffel Tower.

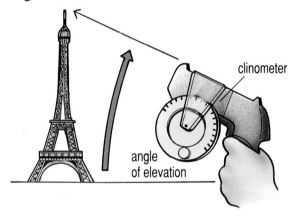
clinometer
angle of elevation

clock ► page 26

clockwise   *adjective*
Clockwise is a movement in the same direction as the hands of a clock.
A ship's bearing is measured in a clockwise direction.

code   *noun*
A code is an arrangement of **symbols** which is used to contain a message. The code may be made up of letters, numbers or a particular graphic.
They used a code to send the message.

coins ► page 28

collinear   *adjective*
Collinear means to be located on the same straight line. Points which are joined by a straight line are collinear points.
In maths class at school, they were shown that the points on the graph axis representing days were collinear.

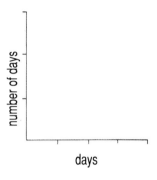
number of days
days

combination   *noun*
A combination is a way of grouping numbers together. In the **set** of numbers 5, 3, 7 and 4, the numbers can be paired in several different combinations 5,3; 5,7; 5,4; 3,7; 3,4; and 7,4. The order of the numbers in a combination does not matter, so 5,3 is the same combination as 3,5.
She made different combinations from the numbers 6, 7, 8 and 9.

commerce   *noun*
Commerce is the buying and selling of goods and services.
The millionaire began in commerce by buying and selling sugar.
commercial   adjective

commission   *noun*
Commission is a fee, or payment, charged by a person who acts on behalf of some other person or company. An insurance agent earns a commission from the insurance company for each policy sold.
The commission was paid to the salesperson after the deal was completed.

clock noun

A clock is an instrument for measuring time. There are two main types of clock. Some have pointers, called hands, which turn slowly and point to numbers around the outside of a circular face. Others, known as digital clocks, show a set of numbers on a screen, or read-out, that change every minute or second.

The hands on the clock showed half past four.

Ancient Egyptian water clock

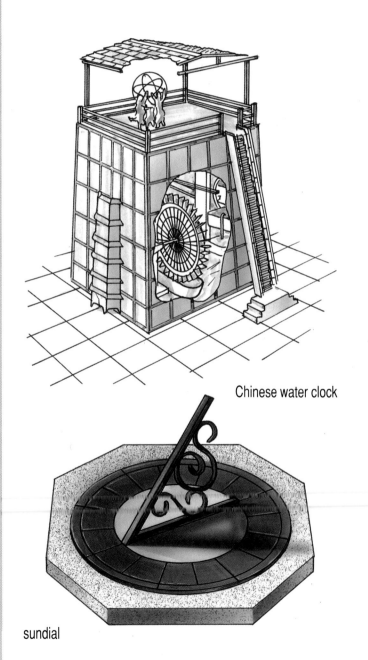

Chinese water clock

candle dial

sundial

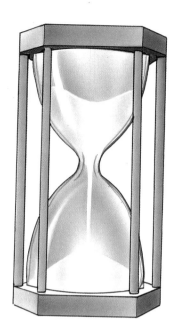

hourglass

pendulum clock

digital watch

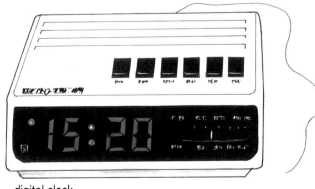

digital clock

coin *noun*

A coin is an item of money made of metal. It is usually of small value only. Money of higher value is issued in the form of **notes**. A coin is usually one of a set. Each coin has a different value relative to a country's main unit of currency. For example, there is a set of seven coins in the United Kingdom varying in value from one pence to 100 pence. One pence is $\frac{1}{100}$ of a pound.

The coins in his pocket were worth only about £1.

Gold stater of Philip the Great of Macedon, 350–336 BC.

Hebrew silver shekel, about 1,900 years old.

This brass coin is elegantly fluted.

Scottish 20 pound piece, James VI

Pierced coins are a feature in some countries.

Maundy money silver penny, George VI.

Retangular coins were used in Japan until the late 1860s.

United Kingdom coins

penny

two pence

five pence

ten pence

twenty pence

fifty pence

one pound

United States coins

cent

five cents (nickel)

ten cents (dime)

twenty-five cents (quarter)

fifty cents (half dollar)

one dollar

commission   *verb*
To commission is to ask someone to do a job for payment.
The company commissioned the artist to paint a picture of the new building.

commodity   *noun*
A commodity is something that is bought or sold. Commodities are often raw materials like wood or iron ore, which will be used to make or manufacture other goods. They may also be raw foods such as sugar cane or coffee beans.
The commodity dealer tried to buy sugar at a very low price.

common-denominator   *noun*
A common-denominator is a **denominator** that is the same for every fraction in a group of fractions. In order to be able to add or subtract a group of fractions they must all have a common-denominator.
She had to work out a common-denominator for each fraction so she could add them together.

communications   *plural noun*
Communications are the various ways of giving or exchanging information. The information can be in the form of words, pictures, electronic signals, signs and symbols or numbers.
Portable telephones allow people to contact each other via an international communications network.

portable telephone

commutative   *adjective*
Commutative means that something can be changed from position to position without altering its value. The commutative law applies to addition or multiplication, because digits can be changed from place to place without altering the answer. For example, $3 + 6$ is the same as $6 + 3$. 8×2 is the same as 2×8.
Subtraction and division are not commutative.

company   *noun*
A company is a type of **business** organization. The people who own it are called the **shareholders** of the company. They usually have **limited liability** for any losses made, in which case the company is called a limited company. A company may be a **private company** or a **public company**.
A company does business in its own name, not in the names of its shareholders.

compass   *noun*
A compass is a device used to locate the north pole. It has a circular face with north, south, east and west marked on it and a magnetic needle that always points to the magnetic north pole.
The needle on the compass showed north-north-east by south-south-west.

compasses   *plural noun*
A pair of compasses is an instrument used to draw a circle or an **arc**. It has two legs which are joined, but can open out to draw curves of any **radius**. One leg has a point at the end and the other has a pencil or pen.
She drew a circle with the compasses.

compensation   *noun*
Compensation is money paid to someone to replace, or make up for, something they have lost. A person who suffers an injury and can no longer work may claim compensation.
He received a year's salary in compensation.
compensate   *verb*

competition   *noun*
Competition occurs when two or more businesses try to win the same customers. Usually each business will try to give customers better **value** than its competitors. It might charge less or make a better product so that people will buy its brand. The opposite of competition is **monopoly**.
The two supermarkets were in competition with each other.
compete   *verb*

complement   *noun*
A complement is the collection of all the things in a **subset** that are not part of a **universal set**. For example, a basket of fruit contains apples, oranges and peaches. The complement of the apples is the **set** of oranges and peaches.
The complement of a set can be worked out on a Venn diagram.

complementary angles   *plural noun*
Complementary angles are two angles that add up to 90°.
Angles that measure 60° and 30° are complementary angles.

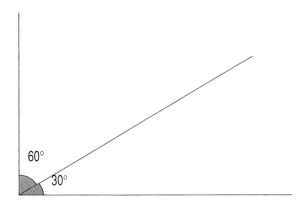

compound interest   *noun*
Compound interest is **interest** calculated on the amount of the original sum of money borrowed, plus any interest which has been added to this sum. It is often paid by banks on savings. A loan for £100 with interest at 10% has a first interest payment of £10. The next interest payment will be calculated on £110, and will be £11.
The bank used compound interest to calculate the money to be paid on the savings account.

computer   *noun*
A computer is an electronic machine that can analyse and store information and make calculations. There are two kinds of computers, analogue computers and digital computers. Most computers are digital computers. The biggest of these are called mainframes, and the smallest are called microcomputers.
A computer was used to analyse and store all the information on this weeks sales in the department store.

concentric   *adjective*
Concentric means to have the same centre. A group of circles which share the same centre are called concentric circles.
When the pebble was thrown into the pond, the ripples formed concentric circles.

cone   *noun*
A cone is a kind of solid **shape** or **polyhedron**. It has a flat, circular bottom and sloping sides. The sides rise up from the **base** and meet at a point at the top, called the **vertex**.
We could see that the top of a rocket was shaped like a cone.
conical   *adjective*

congruent   *adjective*
In **geometry**, congruent describes things that are exactly the same size and shape.
When two congruent triangles are placed on top of each other, they match exactly.

31

constant *adjective*
Constant describes things that do not change. Something that is constant can be used as a measure to show how much other things change. A ruler is a constant length, so it can be used to measure the length of things shorter than itself. The opposite to constant is **variable**.
The temperature at which water freezes is a constant.

construct *verb*
To construct is to make an accurate drawing usually by following mathematical instructions, using special equipment.
She was asked to construct a nautilus, using a ruler and a pair of compasses.
construction *noun*

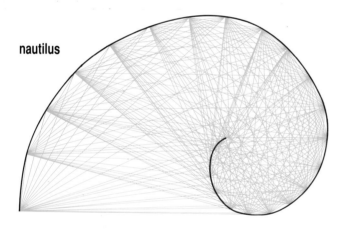

nautilus

consumer *noun*
A consumer is someone who buys things for their own use.
A person who buys envelopes is a consumer of paper goods.

consumer price index *noun*
The consumer price index is a measure of the cost of living. It is based on changes in **retail** prices. Usually, it is worked out by looking at the prices of a selection of goods and services necessary to everyday life. Every so often, new prices are compared to those given when the survey started. This shows how prices have gone up or down.
The consumer price index showed that prices in the country were reducing.

contract *noun*
A contract is a legal agreement between two or more people or **organizations**. Each agrees to do something for the other. Very often, the agreement is that one party will provide goods or services in exchange for payment by the other party.
She entered into a contract to rent the flat .

conversion ▶ page 33

convert *verb*
To convert means to change from one measure to another. The **currency** of one country can be converted into the currency of another. **Conversion** tables can be used to change metric measurements into imperial measurements.
She converted her US dollars into pounds sterling.

co-ordinates *plural noun*
Co-ordinates are a pair of numbers or letters that describe a position on a **graph** or a **map**. The co-ordinates of each point on a graph are shown on the **axes** of the graph.
On a map of the Earth, lines of longitude and latitude are used as co-ordinates.

córdoba *noun*
The córdoba is the currency of Nicaragua. It is made up of 100 centavos.

corporation *noun*
A corporation is a group of people who are given certain legal rights which allow them to act as a single person. It can buy, sell, own and make things as an individual.
The corporation bought a new truck

corresponding angles *plural noun*
Corresponding angles are pairs of angles made when two lines are crossed by another straight line. Each angle in the pair is in the same position. For example, the two bottom left angles are corresponding angles.
The slip road crossed both main roads, making corresponding angles.

conversion *noun*

Conversion is a way of changing a **measurement** into an equivalent measurement using a different system, for example, from metric to imperial.

volume

to convert	multiply by
cubic inches to cubic centimetres	16.387
cubic centimetres to cubic inches	0.061
cubic feet to cubic metres	0.028
cubic metres to cubic feet	35.314
cubic yards to cubic metres	0.765
cubic metres to cubic yards	1.308

length and distance

to convert	multiply by
inches to centimetres	2.540
centimetres to inches	0.394
feet to metres	0.305
metres to feet	3.281
yards to metres	0.914
metres to yards	1.094
miles to kilometres	1.609
kilometres to miles	0.621

liquid volume

to convert	multiply by
cubic inches to litres	0.164
litres to cubic inches	61.027
fluid ounces to millilitres	30.0
millilitres to fluid ounces	0.034
pints (imperial) to litres	0.568
litres to pints (imperial)	1.76
pints (American) to litres	0.47
litres to pints (American)	2.1
gallons (imperial) to litres	4.545
litres to gallons (imperial)	0.22
gallons (American) to litres	3.8
litres to gallons (American)	0.26

surface and area

to convert	multiply by
square inches to square centimetres	6.452
square centimetres to square inches	0.155
square metres to square feet	10.764
square feet to square metres	0.093
square yards to square metres	0.836
square metres to square yards	1.196
square miles to square kilometres	2.589
square kilometres to square miles	0.386
acres to hectares	0.405
hectares to acres	2.471

weight and mass

to convert	multiply by
grains to grams	0.065
grams to grains	15.43
ounces to grams	28.35
grams to ounces	0.035
pounds to grams	453.592
grams to pounds	0.002
pounds to kilograms	0.454
kilograms to pounds	2.205
tons to kilograms	1016.05
kilograms to tons	0.001
metric tonne to short ton	1.1
short ton to metric tonne	0.9

temperature

To convert Celsius into Fahrenheit multiply by 9, divide by 5, and add 32.

To convert Fahrenheit into Celsius subtract 32, multiply by 5, and divide

The figures in this table have been rounded to three decimal points.

cosine *noun*
The cosine is a law in **trigonometry**. It allows the length of one side of a triangle to be calculated, providing the length of the other two sides and the size of the angle between them are known. For a triangle ABC, with sides a, b, and c, the law of cosine is written $C^2 = a^2 + b^2 - 2ab(\cos C)$.
She calculated the length of one side of the triangular shape using the cosine.

cost ► price

counterfeit *adjective*
Counterfeit describes something that is a **forgery**, or fake, but looks very similar to the original. Counterfeit money is pieces of paper that have been printed to look like real bank notes.
The counterfeit money was worthless.

counting *noun*
Counting is a method of finding out how many of a certain item there are. Counting is done by using numerals, or digits, usually on a **base** of 10.
Counting the tins of fruit took the shopkeeper a long time.

cowrie shell *noun*
Cowrie shells are seashells that were once used as **currency** in Africa and Asia. They were used to barter, or exchange, goods.
He gave cowrie shells for the blanket.

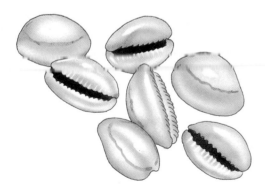

crash *noun*
A crash is the sudden failure of a company or business.
After the crash of the airline company, the investors lost all their money.

credit *noun*
1. Credit is a bookkeeping term which describes a kind of entry in an account. Credit refers to a **profit** or an item of **income** earned by a business. The opposite of credit is **debit**.
A credit entry was made in the sales account to show the value of the sale.
2. Credit is an arrangement a customer can make to pay for goods or services some time after they are bought. The opposite of credit is immediate payment in cash.
Most businesses trade with each other on credit.
3. 'In credit' is a term sometimes used to mean that a person or business is owed money by its bank. The term 'credit balance' is also used to mean this.
The company had a credit balance of £2,000 in the bank.

credit card *noun*
A credit card is a small plastic card that can be used to buy goods. The card contains information about the owner's credit status, which means the amount of money which can be spent on the card. The cardholder has an **account** with the credit card company, and must repay the amount spent over a period of time. The cardholder has to pay the bank back later.
She used her credit card to buy the book.

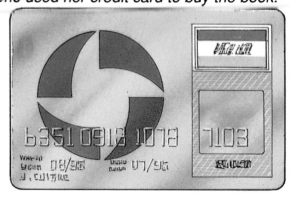

creditor *noun*
A creditor is someone to whom money is owed. The money usually represents supplies or services which have been given.
The company paid its creditors at the end of each month.

cross-section *noun*
A cross-section is the **shape** made when an object is cut through the middle. Cross-sections are often used to show the inside of an engine or other machine.
The cross-section of the motor showed all its working parts as a diagram.

cruzado *noun*
The cruzado is the currency of Brazil. It is divided into 1,000 old cruzeiros.

cube *noun*
1. The cube of a number is the number multiplied by itself three times. The cube of 4 is $4 \times 4 \times 4 = 64$. The cube of 4 is written 4^3, or four to the **power** of three.
The cube of 3 is 27.

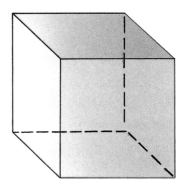

2. A cube is a solid **shape**, or **polyhedron**, with six faces. All the faces are squares of the same size.
She put a cube of sugar in her coffee.

cube root *noun*
A cube root is a number that makes a **cube**. It is worked out by multiplying a number by itself 3 times. The cube root of 64 is 4, because the cube of 4 is $4 \times 4 \times 4 = 64$.
She worked out the cube root of 125 to be 5.

cubic measure *noun*
Cubic measure is a measurement of **volume** in cubic units. The units are **cubes** with equal sides. For example, one cubic metre measures one metre high by one metre wide by one metre deep.
Cubic measure was used to work out how much water put into the swimming pool.

cubit *noun*
A cubit was a unit of length used in ancient times. A cubit was the length of a person's arm from the elbow to the tip of the fingers. It measured about 45 to 55 centimetres.
The carpenter sawed off a piece of wood 2 cubits long.

cuboid *noun*
A cuboid is a solid **shape** or **polyhedron** with six faces. Each face of a cuboid is a **rectangle**. A cube is a special sort of cuboid that has square faces.
All the bricks had a cuboid shape.

currency *noun*
Currency is the money used by a particular country. In the United States the currency is the US dollar, while in Britain the currency is the pound sterling.
When you travel abroad, you need to understand the value of the local currency.

current asset *noun*
A current asset is an asset that is either money already, or one which will be changed into money fairly quickly. Money in a bank account is a current asset. The **stock** held in a shop is also a current asset, because customers will soon buy it for cash.
The company's current assets included cash and stock.

curve *noun*
A curve is a line that bends smoothly. It has no straight parts or sharp corners.
The footballer kicked the ball towards the goal in a curve.
curved *adjective*

customs *plural noun*

Customs are a kind of **duty**, or tax, paid on goods imported into a country. Each country has its own rules about what goods are allowed to be imported, and how much duty is payable. At a port or airport, passengers and their luggage are checked by customs inspectors. In many countries some items, such as illegal drugs and weapons, may be confiscated.

The customs officer asked him to open his suitcase.

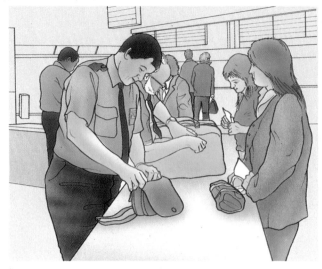

Customs officers may check personal luggage.

Customs boats patrol the harbour ready to solve any problems.

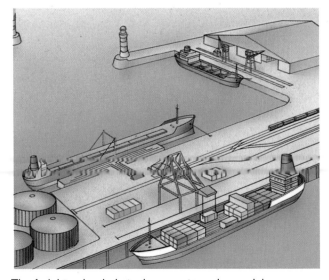

The freight unloaded at a busy port needs special documentation.

Customs officers may make a thorough search if they suspect a traveller of importing something illegally.

Shipments of food must be carefully checked. Many countries don't allow certain products through in an effort to try and stop the spread of pests or diseases.

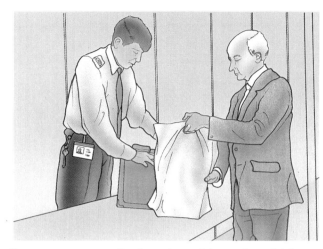

Passengers are usually allowed a certain number of duty-free goods.

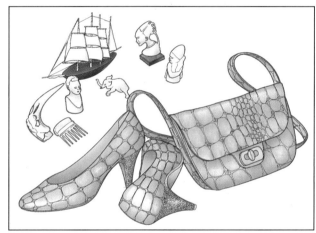

Items made of ivory, crocodile skin and other rare materials can not be brought in to many countries without special permission.

Sniffer dogs are used to detect hidden drugs.

customer *noun*
A customer is someone who buys something.
The customer bought a loaf of bread.

customs ► page 36

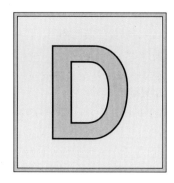

Danish krone (plural **kroner**) *noun*
The Danish krone is the currency of Denmark. The krone has the same value as 100 öre.

data *plural noun*
Data is information or facts. In **finance** the word data is often used to describe information that is put into or stored in a computer. Data can also be withdrawn, or outputted, from a computer in order to supply an **analysis**.
All the data on the financial results was sent to the accountant.

day *noun*
A day is the time it takes for the Earth to turn once on its **axis** as it spins around the Sun. Ordinary clocks count this time as 24 hours. Astronomers measure the length of a day more accurately as 23 hours, 56 minutes. Day is sometimes used to describe the hours of daylight and night for the period of dark.
It took several days for the family to drive down to the coast of Spain for their holiday.

deal *noun*
A deal is an agreement. It is made when one person agrees to do business with another. A deal usually involves buying or selling goods or services.
He made a good deal when he bought the car for only £1,000.
deal *verb*

dealer ► **trader**

debit *noun*
1. In **bookkeeping**, a debit is an entry in an **account**. Debit describes a **loss** or an **expense** made by a business. The opposite of debit is **credit**.
The accountant made a debit entry of $4,000 to show the cost of the new machine.
2. 'In debit' is a term sometimes used to mean that a business owes money to its bank. The term 'debit balance' is also used to mean this.
The company's bank account was in debit to the amount of £1,000.

debt *noun*
A debt is an amount of money that is owed to someone.
His debt of $400 had to be repaid .

debtor *noun*
A debtor is someone who owes money.
The company asked all its debtors to pay their bills.

decade *noun*
A decade is a length, or span, of time lasting 10 years. Ten decades make a **century**.
Parts of Africa have suffered a severe drought for the past decade.

decagon *noun*
A decagon is a plane shape which has 10 sides. In a regular decagon, all the sides are the same length.
An irregular decagon has sides of various lengths.

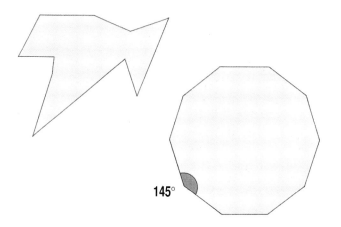

145°

decimal *adjective*
Decimal describes counting in tens. Decimal counting uses the **base** 10. In decimal counting, 1,835 means 1 thousand, 8 hundreds, 3 tens and 5 units. A decimal **currency** is based on multiples and fractions of 10.
The decimal system began when the earliest people began by using their 10 fingers to count.
decimalization *noun*

decrease *verb*
Decrease means to get smaller. In mathematics, totals decrease as numbers or amounts are **subtracted** from them.
The amount of money in her bank decreased after she paid her bills at the end of the month.

deduct *verb*
Deduct means to take away or subtract.
When 9 is deducted from 31, the answer is 22.
deduction *noun*

deficit *noun*
A deficit occurs when the amount of money spent is greater than the amount earned, or received. In accountancy terms, this means that **expenditure** is greater than **income**. Individuals, companies and even countries, can all experience deficits at times. The opposite of deficit is **surplus**.
The company's accounts showed a deficit of £100,000 for the year.

deflation *noun*
Deflation is a reduction in the amount of money available to spend, which makes prices fall. It usually occurs when people, companies or countries cut back on how much money they spend. During periods of deflation, there is less activity in the **economy**.
In times of deflation, prices often start to fall.
deflate *verb*

degree   *noun*
1. A degree is a measurement of an arc or an angle. Degrees can be measured with a **protractor**. A right angle measures 90 degrees. One full **rotation** measures 360 degrees. The **symbol** for degree is °.
The balloon drifted several degrees off course.
2. A degree is a unit of **measurement** for temperature. Water freezes at 0 degrees or 0°, Celsius and boils at 100 degrees, or 100°, Celsius.
The temperature on Thursday was 18 degrees Celsius.

demand   *noun*
Demand is customers wanting to buy, as well as how much they require.
The shop bought extra tea to meet the sudden demand from its customers.

denarius   *noun*
The denarius was a silver coin used in Ancient Rome. The coin was first issued about 211 BC.
A denarius was often stamped with the head of the emperor.

denary   *adjective*
Denary describes something to do with 10. The denary system, in which **units** are sorted into groups of 10, is more often called the **decimal** system, or **base** 10.
In the denary system you count in hundreds, tens and units.

denomination   *noun*
A denomination is the value printed on a coin or bank note, or its **face value**.
The denominations of the coins in his pocket were 10 cents and 50 cents.

denominator   *noun*
The denominator of a fraction is the number written below the line, or **bar**. The denominator shows how many parts a whole number has been split or divided into. It is the number that is being divided into the top number.
The denominator of the fraction $\frac{4}{7}$ is 7.

deposit   *verb*
To deposit means to put money into a bank account. The opposite of deposit is **withdraw**.
He deposited the cheques in his bank account.

deposit   *noun*
A deposit is an amount of money paid in advance to a **supplier**. If a customer wants to buy goods which must be specially ordered, the shop may ask the customer to pay part of the price as a deposit.
A deposit helps to make sure that a customer completes a deal.

depreciate   *verb*
Depreciate means to fall in value, or to become worth less than before. Goods normally depreciate in value as they are used or become older.
The value of the car will depreciate over the coming year.
depreciation   *noun*

depression   *noun*
A depression is a period of time when the **economy** is not very active. During a depression, many people become **unemployed**, factories make fewer products, and people buy fewer goods.
Even for people who still have a job, wages can fall during a depression.

40

design ► pattern

deutsche mark   *noun*
The deutsche mark, sometimes spelled deutschmark, is the currency of Germany. The deutsche mark is made up of 100 pfennigs.

devalue   *verb*
Devalue means to make something worth less than it used to be.
The government decided to devalue its currency compared with other currencies.
devaluation   *noun*

diagonal   *noun*
A diagonal is a straight line that can be drawn across a shape from one corner to another. If you draw a diagonal across a square, you will cut the square into two equal triangles.
She folded the silk square diagonally.

diagram   *noun*
A diagram is a drawing used to illustrate a problem in **geometry**. It can also describe a plan or a **chart** which shows a simple outline of a more difficult **construction**.
He drew a diagram to show how far the planets are from the Sun.
diagrammatic   *adjective*

diameter   *noun*
The diameter of a circle is the distance across the middle through its centre point. This distance is always twice the length of the **radius**.
The diameter of the water lily was more than one and a half metres.

diamond ► rhombus

die (plural **dice**)   *noun*
A die is a small **polyhedron**, usually with a group of dots or a number stamped on each of its faces. When the die is a cube, the number of dots stands for the numbers 1, 2, 3, 4, 5 and 6.
A die is thrown in many board games to decide how many places a player is allowed to move.

difference   *noun*
Difference is another way to describe subtraction. The difference between two numbers is the answer obtained when one number is subtracted from a second which has a higher value.
If 3 is subtracted from 7, the difference is 4.

digit   *noun*
A digit is a single number. In the decimal system of counting, the digits are 0, 1, 2, 3, 4, 5, 6, 7, 8, 9. Digits can be used to write larger numbers. The number 7,428 has four digits, 7, 4, 2 and 8. In the **binary** system, there are only two digits, 0 and 1.
Another name for a digit is a numeral.

digital   *adjective*
Digital describes anything that uses **digits** to display information.
My digital telephone number is 8723-4660.

dime   *noun*
A dime is a coin used in the United States of America and Canada. It is worth 10 cents.

dimension ► page 42

dimension   *noun*

Dimension is a **measurement**, such as length or thickness. A straight piece of thread has one dimension, that of length. A flat piece of paper has two dimensions, its length and its width. A box has three dimensions, length, width and height.
The dimensions of the page are 270 millimetres high and 217 millimetres wide.

This boat has been drawn to show two dimensions, length and height.

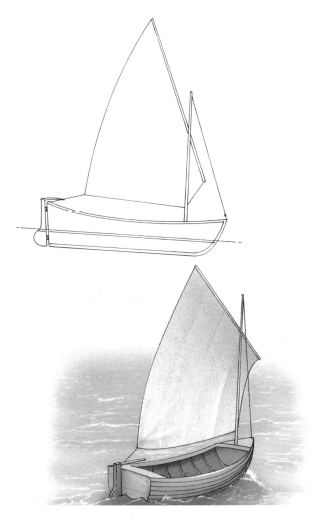

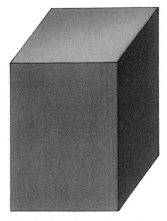

A square has two dimensions, length and height.

A cube has three dimensions length, width and height.

Artists use perspective to create an impression of depth. In this drawing the boat appears to have three dimensions.

Artists can play with perspective to create shapes on paper that couldn't really exist as objects.

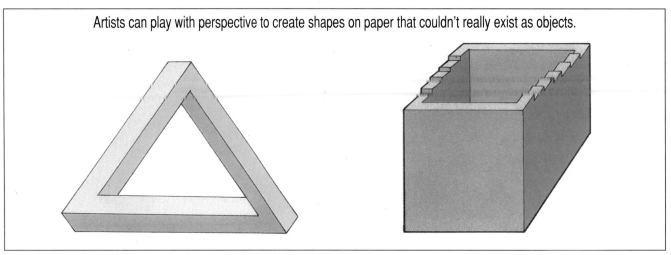

42

dinar *noun*
The dinar is the unit of currency in several countries of the world, such as Algeria, Jordan and Tunisia. In Algeria, one dinar is worth 100 centimes. In Jordan it is divided into 1,000 fils and in Tunisia, 1,000 millimes.

direct proportion *noun*
Direct proportion shows how one measurement changes at the same rate as another. For example, if two kilograms of fruit cost 3 dollars then four kilograms will cost 6 dollars. The weight of the fruit and the cost of the fruit double in direct proportion.
The wall grew taller in direct proportion to the number of bricks used.

directed number *noun*
A directed number is a number that has either a plus or minus sign before it. A plus or + number is known as a positive number. A minus or − number is called a negative number.
Negative numbers are directed numbers.

director *noun*
A director is a person who manages the business of a company. A company usually has several directors, who together make up the **board of directors**. The directors are chosen by the people who own the company, the **shareholders**.
The directors decided that the company should begin to make a new product.

dirham *noun*
The dirham is the unit of currency in Morocco and in the United Arab Emirates. In Morocco, one dirham is equal in value to 100 centimes. In the United Arab Emirates, one dirham is made up of 100 fils.

discount *noun*
A discount is a reduction in price. Goods are normally sold at a discount to encourage people to buy them.
The full price of the book was $20 but with the discount it was only $17.

distance *noun*
Distance is the **measurement** between two places or points. The metric units of distance, or **length**, are kilometres, metres, centimetres and millimetres. In imperial measurements, distance is calculated in miles, yards, feet and inches.
She measured the distance between the two trees with long strides.

dividend *noun*
A dividend is money paid by a company to its shareholders.
Dividends are often paid yearly.

divisibility test ► page 44

division *noun*
Division is a kind of arithmetic. The sign for division is ÷. If there are twelve biscuits and six people, then division is used to work out that each person can have two biscuits. This can be written $12 \div 6 = 2$.
Division is used to share out things equally.
divide *verb*

dodecagon *noun*
A dodecagon is a twelve-sided flat, or **plane shape**. It is a type of **polygon**.
The sides of a regular dodecagon are the same length.

dodecahedron *noun*
A dodecahedron is a **solid shape** with 12 faces. It is a type of **polyhedron**.
The faces of a dodecahedron have five equal sides.

dollar ► page 45

domino *noun*
A domino is a rectangle made by joining two squares together. The long side of a domino is twice the length of the short side. A domino is a **polyomino** that can be used in **tessellation**.
A popular counting and number matching game has the name, dominoes.

divisibility test   *noun*

A divisibility test is used to find out whether or not one number is divisible, or can be divided equally, by another number. For instance, a number is divisible by 2 if it ends in 0, 2, 4, 6 or 8. If you apply the divisibility test to the number 146 you can see that it ends in 6, so it must be divisible by 2.
We practised using divisibility tests in class today.

by 2 if it ends in 0, 2, 4, 6, or 8

A number is divisible:

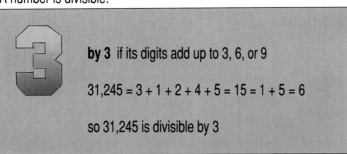
by 3 if its digits add up to 3, 6, or 9

$31{,}245 = 3 + 1 + 2 + 4 + 5 = 15 = 1 + 5 = 6$

so 31,245 is divisible by 3

by 4 if its last two digits are divisible by 4 the whole number is divisible by 4

16 will divide exactly by 4 so the whole number 2,716 is divisible by 4

$2{,}716 \div 4 = 679$

by 5 if it ends in 0 or 5

by 6 if it is an even number and divisible by 3

by 9 if its digits add up to 9

$567 = 5 + 6 + 7 = 18 = 1 + 8 = 9$

$567 \div 9 = 63$

by 11 if it is a three digit number whose two outside digits added equal the one in the middle

132   $1 + 2 = 3$ 3 is in the middle
198   $1 + 8 = 9$ 9 is in the middle
091   $0 + 1 = 9$ 9 is in the middle

$132 \div 11 = 12$
$198 \div 11 = 18$
$891 \div 11 = 81$

by 10 if it ends in 0

dollar *noun*

The dollar is the unit of currency of many countries of the world, including Australia, the Bahamas, Barbados, Belize, Canada, Dominica, Fiji, Guyana, Hong Kong, Jamaica, Liberia, New Zealand, Singapore, Taiwan, Trinidad and Tobago, the United States of America and Zimbabwe. The word came from the English for the German 'thaler' coin. The sign for dollar is $.

Many countries issue their dollar in both paper notes and coins.

The American dollar became the official currency of the United States of America in 1792. It is issued as $1, $2, $5, $10, $20 and $50 notes.

The Canadian dollar is circulated as a $1 coin, and $1, $2, $5, $10, $20, $50, $100 and $1,000 paper notes.

The Singapore dollar is issued as a $1 coin and $1, $5, $10, $20, $50, $100, $1,000 and $10,000 paper notes.

Hong Kong dollar notes of $10 and over are issued by the Hong Kong and Shanghai Banking Corporation and the Standard Chartered Bank.

The Singapore $20 note features a yellow-breasted sunbird.

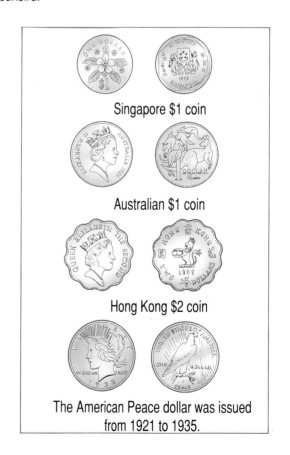

Singapore $1 coin

Australian $1 coin

Hong Kong $2 coin

The American Peace dollar was issued from 1921 to 1935.

double *verb*
To double means to multiply by two.
After taking a rest, he set off at double his previous speed.

doubloon *noun*
A doubloon was a gold coin that was used in Spain until the 1800s.
In the past, pirates raided Spanish merchant ships to steal their treasure of doubloons.

Dow Jones Index *noun*
The Dow Jones Index is an **index**, or number, which indicates the present level of the price of shares on the New York Stock Exchange, compared to past levels. It changes continually as new business is done.
The Dow Jones Index rose on Tuesday as more people bought shares.

dozen *noun*
A dozen is another word for 12. Goods are often sold by the dozen.
The chef beat together a dozen eggs to make the omelettes.

drachma *noun*
The drachma is the currency of Greece.
One drachma is worth 100 lepta.

draft *noun*
A draft is a kind of **cheque**. A bank draft is used to transfer money from one bank account to another. The accounts can be in the same bank or in different banks.
The bank issued a draft so that he could pay for the boat.

ducat *noun*
Ducat was the name given to several different kinds of gold or silver coin that were once used in Europe.

duodecimal *adjective*
A duodecimal system is a number system in which the **base** is 12. In base 12, the number 34 shows that there are four units and three 12s. There are 12 symbols in base 12, so 0–9 are used plus two others to represent 10 and 11. Usually, letter symbols are used such as x or y.
English currency used a 12 pence to the shilling duodecimal system before decimalization.

duty *noun*
Duty is a kind of **tax**. When goods are imported, exported or sold, an extra amount of money sometimes has to be paid to the government. This extra payment is called the duty.
He paid duty on the jewellery he brought into the country.

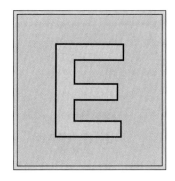

eagle   *noun*
An eagle was a gold coin once used in the United States of America. It was worth 10 dollars.

earnings   *plural noun*
Earnings are the amount of money that is paid to a person for doing a job. Earnings can also mean the amount of **profit** made.
He spent his earnings in July on a holiday in Greece.
earn   *verb*

economics   *noun*
Economics is the study of all the ways in which goods and services are produced, distributed and used. It covers such subjects as the **circulation** of money, **imports** and **exports**, **earnings**, **employment** and **tax**, **inflation** and **deflation** and many more.
The analysis of a country's imports and exports is an important part of economics.

economist   *noun*
An economist is a person who studies **economics**. Economists try to **forecast** what will happen to a country's economy.
The economists made different forecasts of next year's rate of inflation.

economy   *noun*
1. The economy of a country is everything to do with the way it produces things and sells them. It also covers its system of **banks**, **building societies**, **stock exchange** and other financial institutions. One task of a government is to manage the country's economy so that the people grow wealthier.
The government managed the economy well.
2. Economy also means trying to keep costs and expenditure as low as possible.
The directors insisted on strict economy in running the business.

ecu   *noun*
An ecu was a gold or silver coin that was once used in France.

ECU   *abbreviation*
ECU is short for European Currency Unit. It is a currency of the European Community. Businesses or individuals can pay for goods or take out loans in ECUs. It is stamped with 12 emblems to represent the 12 countries which are members of the European Community.
Some banks sell traveller's cheques in ECUs.

Egyptian pound   *noun*
The Egyptian pound is the currency of Egypt. An Egyptian pound is worth 100 piastres or 1,000 millièmes.

element ► member

47

ellipse   *noun*
An ellipse is a regular **oval**. It is the **shape** seen when a cone is cut off on a diagonal.

An ellipse has the shape of a squashed circle.
elliptical   *adjective*

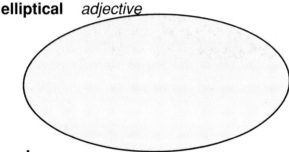

embargo   *noun*
An embargo is a ban on importing or exporting goods. For example, one country might ban the export of weapons to a country that it is not friendly with.
There is an embargo on crocodile skins.

employee   *noun*
An employee is a person who is paid a wage or salary by another person, called the **employer**. Employees are paid to do a job, usually on a regular basis, over a period of time.
The employee was paid to deliver cheese to all the stores.

employer   *noun*
An employer is someone who pays a wage or salary to another person to do a job. An employer can be an individual or a business.
The employer needed another shop assistant.

empty set   *noun*
An empty set is a **set** which has no **members**, or **elements**, in it. The **symbol** to show an empty set is {} or Ø.
In a boy's school, the set of girl pupils would be an empty set.

entrepreneur   *noun*
An entrepreneur is a business person.
An entrepreneur will sometimes take a risk to make a profit.

equal   *adjective*
Equal means the same in quantity, value, size or degree. Two things that balance are equal. The equal sign, =, is used between two sides of an equation.
One pound sterling is equal to 100 pence.

equation   *noun*
An equation is a **statement** that shows that two things are equal. An equation is made up of two quantities with an equal, or =, sign between them. For example, $7 = 5 + 2$ shows that the sum of 5 and 2 is equal to 7.
Equations are often used in algebra.

equilateral triangle   *noun*
An equilateral triangle is a triangle with all three sides the same length. The interior angles of an equilateral triangle are 60°.
Equilateral triangles can be used in tessellation.

equity   *noun*
Equity is a **share** in a particular business or in an **asset**. Each owner of a business will hold some equity in it. Equity is measured by the number of ordinary shares a shareholder owns.
Some of the equity in the business was lost when the shares were sold.

escudo   *noun*
The escudo is the currency of Portugal. An escudo is worth 100 centavos. A thousand escudos is called a conto.

estimate   *noun*
An estimate is a guess at the size of something. An estimate often comes out a little larger or a little smaller than the actual size. A builder may need to estimate how many bricks will be needed in order to build a house.
The driver estimated that the journey would take two hours.

eurodollars   *noun*
Eurodollars are United States dollars held in banks outside the United States, especially in European banks.
The United States of America uses Eurodollars to pay for things bought from foreign countries.

European Currency Unit ► ECU

even number   *noun*
An even number is a number that can be divided exactly by 2. For example, 4, 6, 10 and 24 are all even numbers. A number that cannot be divided exactly by two is an odd number. An even number always ends in 0, 2, 4, 6 or 8.
The sum of two even numbers is always an even number.

exchange rate   *noun*
The exchange rate shows how much one **currency** is worth when it is changed into another currency. The exchange rates between different currencies are changing all the time.
At the present exchange rate, he will get more than 1,000 lire for one dollar.

expenditure   *noun*
Expenditure is money that has been spent. Companies or people in business need to spend money to buy the supplies they need. A printer will buy paper and ink, and possibly a new printing machine as part of the necessary expenditure of the business.
The company's expenditure rose when the new machines were ordered.

expenses   *plural noun*
Expenses are amounts of money that a company or person needs to spend to carry out work.
The saleswoman needed to hire a car, so the company paid her expenses.

expensive   *adjective*
Expensive describes things that cost a great deal of money to buy.
The dress created by the French fashion designer was very expensive.

exponent   *noun*
An exponent is a number written to the right and above another number. It shows how many times a number is multiplied by itself. Another word for exponent is **power**.
Six to the exponent three is 6^3.

exponential growth ► page 50

export   *verb*
To export means to sell goods or services to customers in other countries. The opposite of export is **import**.
The tea was exported from Sri Lanka to Great Britain.

exports   *plural noun*
Exports are goods or services sold to customers in other countries. The opposite of exports is **imports**.
If exports increase, the wealth of the country will grow.

exponential growth *noun*

Exponential growth, also known as geometric progression, is a rapid increase in number. The first number is multiplied by a value. The result is multiplied by the same value, and so is each result after that. The equations $2 \times 3 = 6$, $6 \times 3 = 18$, $18 \times 3 = 54$ show that {6, 18, 54} is an example of exponential growth.

Exponential growth was seen in the increase of the number of fruit flies this summer.

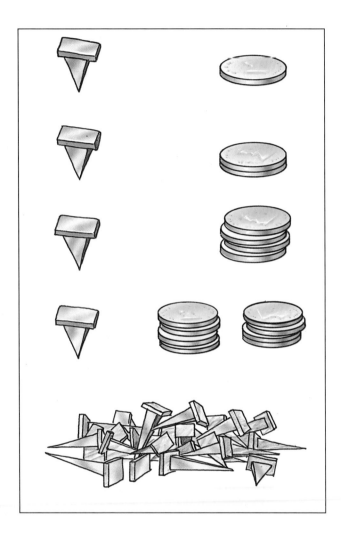

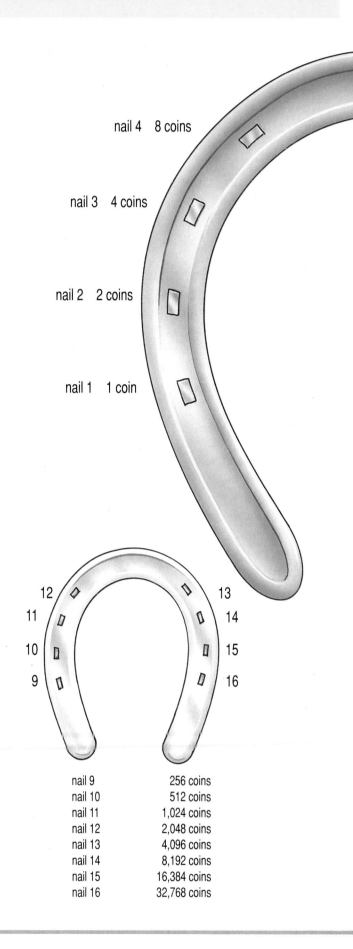

nail 4   8 coins

nail 3   4 coins

nail 2   2 coins

nail 1   1 coin

12 11 10 9   13 14 15 16

nail 9	256 coins
nail 10	512 coins
nail 11	1,024 coins
nail 12	2,048 coins
nail 13	4,096 coins
nail 14	8,192 coins
nail 15	16,384 coins
nail 16	32,768 coins

There is an old story about a blacksmith who agreed to shoe a horse for one gold coin for the first nail, two for the second, four for the third, eight for the fourth and so on. He was due 4,294,967,295 gold coins by the time the 32 nails were in place.

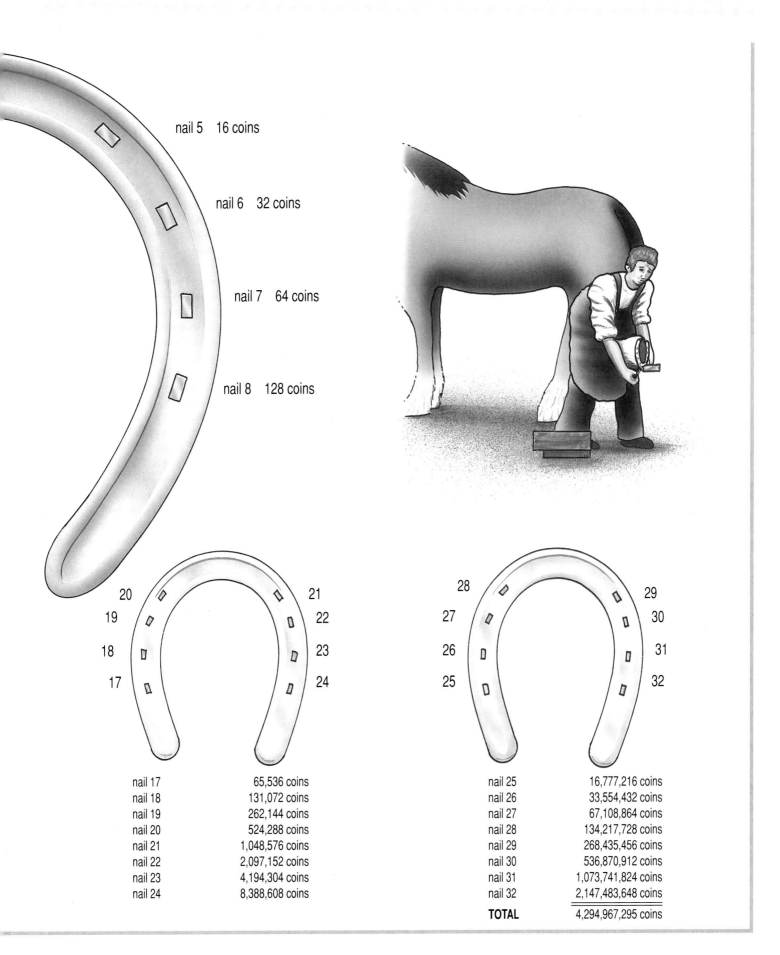

nail 5   16 coins

nail 6   32 coins

nail 7   64 coins

nail 8   128 coins

20   21
19   22
18   23
17   24

28   29
27   30
26   31
25   32

nail 17	65,536 coins
nail 18	131,072 coins
nail 19	262,144 coins
nail 20	524,288 coins
nail 21	1,048,576 coins
nail 22	2,097,152 coins
nail 23	4,194,304 coins
nail 24	8,388,608 coins

nail 25	16,777,216 coins
nail 26	33,554,432 coins
nail 27	67,108,864 coins
nail 28	134,217,728 coins
nail 29	268,435,456 coins
nail 30	536,870,912 coins
nail 31	1,073,741,824 coins
nail 32	2,147,483,648 coins
TOTAL	4,294,967,295 coins

exterior angles *plural noun*
The exterior angles of a **polygon** are the angles which each side of its shape turns through. At each point of a polygon, the **exterior angles** and the **interior angles** are **supplementary angles**.
The exterior angles of a hexagon are all equal to 60°.

face value *noun*
The face value of something, such as a bank note, is the value that is printed on it. The face value of a five-pound note is five pounds. But something may be worth more than its face value. The value of a **share** may change, so it is possible to pay more for it than its face value.
The broker had to pay double the face value of the shares.

factor *noun*
A factor is a number that divides equally into another number with nothing left over. The number 4 is a factor of 12 because there are exactly three 4s in 12, so 12 can be divided by 4 exactly.
She worked out that the factors of 6 are 1, 2, 3 and 6 as each of these numbers divide equally into it.

factorization *noun*
Factorization is the process of finding all of the numbers that are a factor of a particular number. The **factors** of 12 are 1, 2, 3, 4, 6 and 12 because 12 can be divided by any of these numbers.
He did the factorization of 14 to get 1, 2, 7 and 14.
factorize verb

factory *noun*
A factory is a building where a company makes the goods that it will sell. The factory houses the production equipment and the employees who work there.
The cars were made on an assembly line in a large factory.

Fahrenheit *noun*
Fahrenheit is a **scale** used to measure temperature in degrees. The freezing point of water is a **constant** 32 degrees on the Fahrenheit scale (32°F). The boiling point of water is 212 degrees Fahrenheit (212°F). Today, the **Celsius** scale is more widely used.
The mid-day temperature was 40 degrees Fahrenheit.

fare *noun*
A fare is a payment you make to travel on a vehicle such as a bus, train or plane.
He bought the ticket with some coins.

farthing *noun*
A farthing is a small coin that was once used in Britain. When it was used it was worth a quarter of an old penny.

fee *noun*
A fee is an amount charged by a person who performs a **service** for someone else. Accountants, lawyers, consultants, advertising agencies and other service businesses charge fees to their clients.
The company paid a fee to a consultant for advice on which computer to buy.

figure *noun*
1. A figure is a **digit**. The figures used in the **decimal system** of counting are 0, 1, 2, 3, 4, 5, 6, 7, 8 and 9.
The number 2,386 has four figures.
2. Figure is another word for **shape**. Figures can be **planes** or **solids**.
A circle is a plane figure and a sphere is a solid figure.

finance *noun*
Finance is the money used by a business in its day to day work. The term finance can also refer to sums of money set aside for particular projects.
The company was seeking finance for a building project.
financial *adjective*

finance *verb*
To finance means to provide the money needed to develop or to create a business. A government may decide to help finance projects in regions where unemployment is high.
The large company financed the building of a new liner.

Financial Times Index *noun*
The Financial Times Index is an **index**, or number, which indicates the present level of the price of shares at the London Stock Exchange, compared with past levels. It is based on the share prices of 30 leading British industries.
The Financial Times Index dropped as share prices went down and people sold their shares quickly.

financier *noun*
A financier is a person who handles large sums of money. The financier provides money to invest in projects which it is believed will be successful and which will earn **interest** plus **profit** on the money provided.
The financier provided £300,000 for the new project.

firm *noun*

A firm is a general word meaning a business. It usually refers to a business which is not set up as a **limited company**. A firm of accountants or lawyers is not a company because the owners, or the partners, of the firm do not have limited liability.
There were seven partners in the firm of accountants.

fiscal *adjective*

Fiscal describes the systems a government puts in place to control how it gets money from the public. Governments do this by setting **taxes** and **duties** which the public has to pay.
The government is planning new fiscal policies to increase taxes.

fixed asset *noun*

A fixed asset is an asset which a business owns over a long period of time and which it uses to earn **profits**. A delivery van is a fixed asset because it may last for several years before it is replaced. An office building or a factory are also fixed assets since they may have a very long life.
The company's fixed assets include a factory building.

florin *noun*

1. A florin was a gold coin used in Florence in the thirteenth **century**.
2. Florin was the name of various coins once used in several European countries and in South Africa.

flowchart ► page 56

foot (plural **feet**) *noun*

A foot is an **imperial** measure of length. One foot is equal to 12 inches. The abbreviation for foot or feet is ft. The symbol is ' and is written after a measurement like this, 6'.
The sitting room in the old house measured 17 feet by 20 feet.

forecast *verb*

Forecast is a word that means to predict, or **estimate**, what will happen in the future. In business, companies try to forecast how much they will earn in the coming year. They do this by estimating how much they will sell to customers and how high their costs will be.
They forecast that they would increase their profit next year.

foreign exchange *noun*

Foreign exchange is the buying and selling of **currencies**. **Exchange rates** change from day to day so it is possible to buy US dollars for one price on Monday, and sell them for a better price on Friday. A **profit** is made if this works. However, if exchange rates change in a different direction than expected, a **loss** is made instead.
The company made a profit on foreign exchange this year.

forger *noun*

A forger is someone who makes fake or **counterfeit** money. It is also someone who copies something of value in such a way that it can be mistaken for the original.
The forger printed a pile of new notes.

forgery *noun*

A forgery is something that is made to look like something else. Forgeries are often pieces of paper, such as money and letters, that are substituted for the real thing and used to cheat people.
The forgery was discovered and the criminal sent to jail.

forint *noun*
The forint is the currency of Hungary. A forint is made up of 100 fillér.

formula (plural **formulas** or **formulae**) *noun*
In **algebra**, a formula is a rule, or **equation,** using symbols. The formula for finding the circumference of a circle is 2πr (2 × π × r). Other formulas can be made up to fit any problem as long as it is known what each symbol stands for.
He worked out the area of the playing field using the correct formula.

Fort Knox *noun*
Fort Knox is a place in Kentucky, in the United States of America, where the government keeps its stores of valuable gold **bullion**. The bullion is heavily guarded.
Tourists in the United States of America often visit famous Fort Knox.

fractal *noun*
A fractal is a complex geometric pattern representing a mathematical **equation**.
Fractal shapes can be generated by computer.

fraction ► page 58

franc *noun*
The franc is the unit of currency of several countries, such as Belgium, Cameroon, France, Niger, Senegal and Switzerland. The franc is made up of 100 centimes.

fraud *noun*
Fraud means cheating. If the manager of a company makes false accounts so that money can be kept for personal use, this is a case of fraud. Fraud is a crime.
He went to prison for fraud.
defraud *verb*

free trade *noun*
Free trade is **trade** across the borders of countries that is not interfered with by governments. It refers to goods that are exchanged without **taxes** or **duty** being added.
The government allowed foreign goods to enter the country under its free trade policy.

freezing point *noun*
Freezing point is the temperature at which a liquid turns into a solid. Because the freezing point of water is a **constant**, we can use it to help measure temperature.
The freezing point of water is 0° Celsius.

French franc *noun*
The French franc is the currency of France. It is divided into 100 centimes.

function *noun*
A function is the relationship of one **variable** to another. The quantity or **value** of one variable is related to that of the other. The acceleration of a car is a function. The variables are the car's speed and the time it started to move. If either one of those numbers change, the other is affected.
He drew a line showing changes to the functions of size and price on a graph.

fund *noun*
1. A fund is an amount of money set aside for a particular purpose. Money is often put into a fund in order to save it for a future purchase.
The office built up a coffee fund of £100.
2. Funds is a term sometimes used in a very general way to mean money.
She had enough funds to buy a new coat.

flowchart *noun*

A flowchart is a diagram that shows the steps in solving a problem. The steps are ordered in sequence, so that each must be resolved before proceeding to the next. Flowcharts are generally used as visual aids to help make complex problems easier to analyze.

He used a flowchart to show each step of the solution.

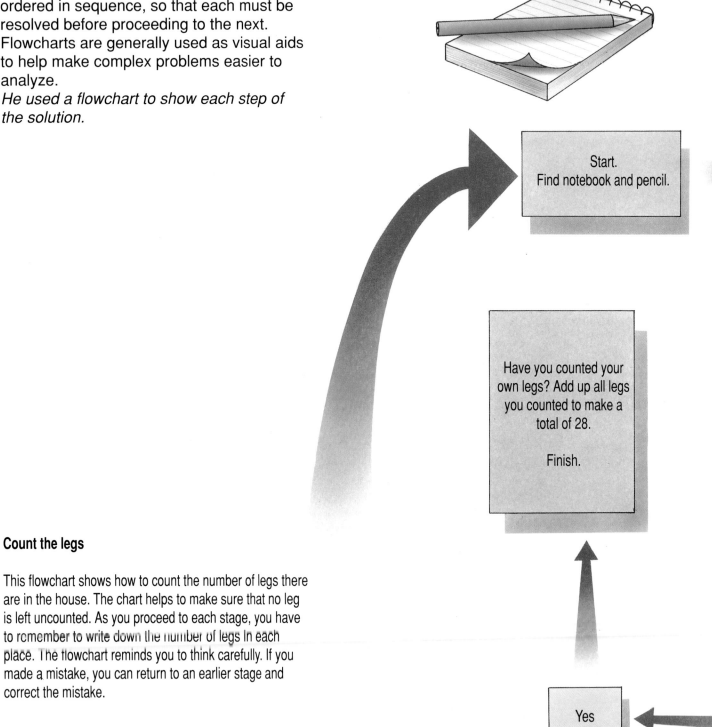

Count the legs

This flowchart shows how to count the number of legs there are in the house. The chart helps to make sure that no leg is left uncounted. As you proceed to each stage, you have to remember to write down the number of legs in each place. The flowchart reminds you to think carefully. If you made a mistake, you can return to an earlier stage and correct the mistake.

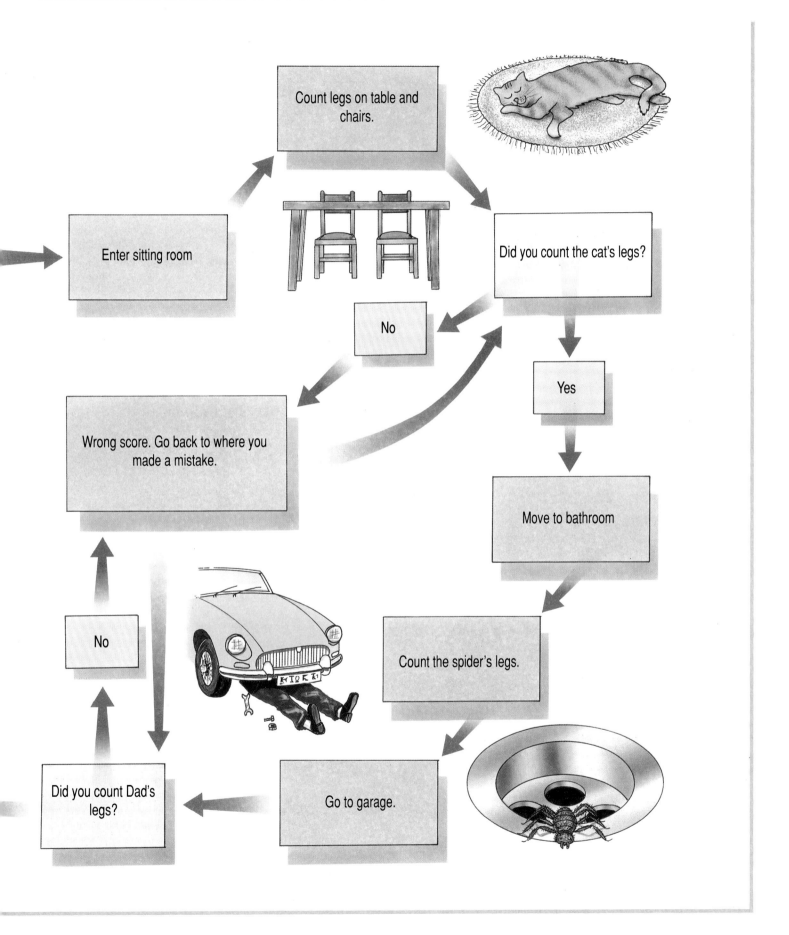

Enter sitting room

Count legs on table and chairs.

Did you count the cat's legs?

No

Yes

Wrong score. Go back to where you made a mistake.

Move to bathroom

Count the spider's legs.

No

Go to garage.

Did you count Dad's legs?

fraction *noun*

A fraction is a small part of something. Fractions are usually written as two numbers separated by a line, or **bar**. The **denominator** is the number below the bar. It is the number of equal parts something has been divided into. The top number, or **numerator**, tells how many of those parts there are. Fractions can also be written, or expressed, as a **ratio**.

He divided the cake into six fractions.

Equivalent fractions can have different numerators and denominators. For example, one half can be written $\frac{6}{12}$, $\frac{5}{10}$, $\frac{4}{8}$, $\frac{2}{4}$, or $\frac{1}{2}$. When one half is written as $\frac{1}{2}$ the fraction is using the lowest common denominator.

$$\frac{1}{4} \; + \; \frac{1}{4} \; = \; \frac{1}{2}$$

$$\frac{3}{4} \; - \; \frac{1}{4} \; = \; \frac{1}{2}$$

$$\frac{1}{4} \; \times \; 2 \; = \; \frac{1}{2}$$

$$\frac{1}{4} \; \div \; 2 \; = \; \frac{1}{8}$$

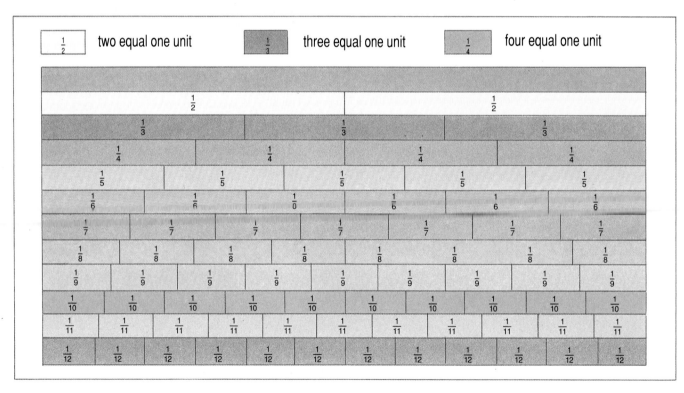

$\frac{1}{2}$ two equal one unit $\frac{1}{3}$ three equal one unit $\frac{1}{4}$ four equal one unit

fund   *verb*
To fund a project means the same as to
finance a project.
*The government decided to fund a housing
project.*

gain ▶ **profit**

gallon   *noun*
A gallon is an **imperial measure** of liquid.
It is divided into four quarts. A gallon is equal
to 4.546 litres in the **metric system**.
*The car needed 12 gallons of petrol to fill up
the tank.*

gamble   *verb*
To gamble means to take a risk to make
money. Some people gamble by playing
games at a **casino** or by placing racing **bets**
on horses. A company may gamble that they
will sell a large number of their new
products.
*At a race course, racegoers gamble on the
winning horse.*

geometry   *noun*
Geometry is the study of lines, shapes and
angles. Geometry is used to work out the
angles in a **polygon**, but **trigonometry** is
used when the polygon is a triangle.
The pupils studied geometry.
geometric   *adjective*

gold *noun*

Gold is a rare, yellow metal. For thousands of years, people have valued gold because it is not affected by air or water, so it never rusts or grows dull. It is also soft, and easy to work into different shapes. Gold was once used to make coins, but is now used mainly in jewellery and in industry.

She bought a necklace made of gold.

Lumps of gold called nuggets are sometimes found.

Gold may be dug from mines underground.

Some grains of gold are found in the bed of a river. The gold is separated from the mud and gravel by panning.

Gold is also obtained by open-cast mining.

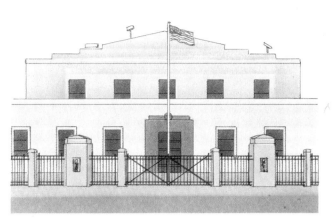

Much of the gold owned by the government of the United States of America is kept at Fort Knox, Kentucky.

Some countries hold large reserves of gold bullion.

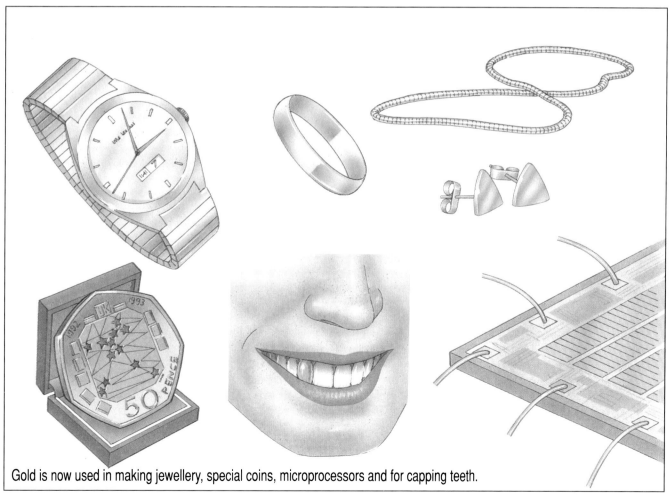

Gold is now used in making jewellery, special coins, microprocessors and for capping teeth.

gilt   *noun*
A gilt is a short way of describing a gilt-edged security. In Britain, a gilt is an **investment** in a **stock** issued by the government. This carries little or no risk, because the government will repay the **loan** and in the meantime it will pay **interest**. In the United States of America, a gilt-edged security has a different meaning. It is an **investment** in a company with a very reliable financial record.
The government sold gilts in its new transport scheme.

gold ► page 60

golden ratio ► **ratio**

gold standard   *noun*
The gold standard was a system used in the past to fix the value of a currency. For example, if a coin or note was given a value equal to the price of an ounce of gold, the value of that **currency** would change whenever the value of gold changed.
The gold standard is no longer in use since it caused many difficulties.

goldsmith   *noun*
A goldsmith is a person who makes objects out of gold.
The goldsmith made a gold ring.

goods   *plural noun*
Goods are products that can be bought. Food, clothes, books, furniture and electrical machines are all examples of goods that can be bought in shops. Another word for goods is merchandise.
The shop ordered more goods.

gradient ► **slope**

gram   *noun*
A gram is a unit of weight in the **metric system**. There are 1,000 grams in a kilogram.
The packet of biscuits weighed 125 grams.

graph ► page 63

greenback   *noun*
Greenback is a familiar term used to describe the paper money used in the United States of America.

grid   *noun*
A grid is a pattern of **parallel** lines which is often also called a lattice. **Graphs** are often recorded on a grid. Many maps include a grid which is given **co-ordinates**.
He charted the team's scores on a grid.

gross   *noun*
Gross means complete, or with nothing taken out. It can refer to wages earned before any **tax** or other **deductions** are taken out or to **profit** that a company makes before it is taxed or pays its **expenses**.
The company reported a gross profit.

Gross National Product   *noun*
Gross National Product is the total value in money of all the **goods** and **services** produced by a country during a particular period of time. It includes the **net** income it receives from other countries. It is used to work out whether the **economy** of a country is improving or declining.
The Gross National Product of the country was calculated in billions of dollars.

gross profit   *noun*
Gross profit is the amount of money that is earned from selling goods after the money it cost to buy them has been deducted.
The company earned a gross profit of a million dollars last year from selling lamps.

guarantee   *noun*
A guarantee is a promise to do something, or that goods are of a certain quality. Electrical equipment often has a guarantee that it will be repaired or replaced if it does not work properly.
The guarantee on the fridge lasted for a year.

graph *noun*

A graph is a diagram normally drawn on a **grid**, which shows how two or more sets of information are related. A line graph has two axes. The vertical **axis** points upwards and the horizontal axis is drawn from left to right. Another kind of graph, a bar chart, is made up of vertical or horizontal bars. A block graph compares information in blocks of different size.

A graph was drawn to show the number of refugees arriving in the camp each day.

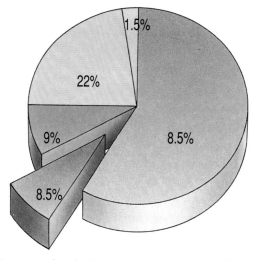

Each sector of a pie chart gives a percentage of the whole.

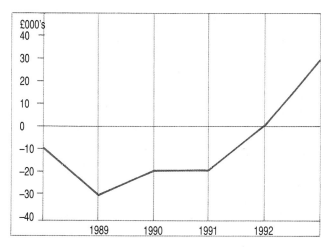

This line graph is used to show a current account balance. One axis is for years, the other for money.

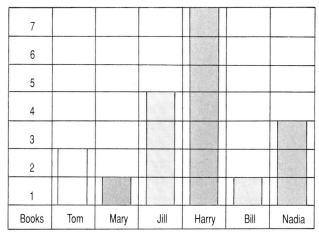

An easy to read bar graph shows how many books each child has read in a given space of time.

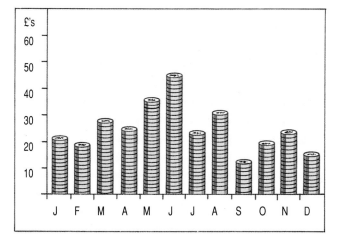

This pictograph uses pictures of money to show savings over a 12 month period.

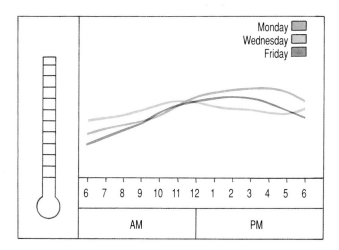

A multiple graph like this can be used to show different sets of information. Here temperature per hour is shown for three days.

63

guilder *noun*
The guilder is the currency of The Netherlands. A guilder is made up of 100 cents.

guinea *noun*
A guinea was a gold coin once used in Britain. It was worth 21 shillings, or 1.05 pounds.

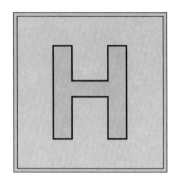

half (plural **halves**) *noun*
A half is a **fraction**. It is one part of a number or an object after it has been divided into two equal parts. Half of six is three because there are two threes in six.
A semicircle is half a circle.

half dollar *noun*
A half dollar is a coin used in the United States of America and Canada. It is worth 50 cents.

heads and tails *noun*
Heads and tails are the faces of a coin. In many countries, one side of each coin has a head on it. This side is called heads. The other side is called tails. Heads and tails are sometimes used to decide who will start a sporting event. One player chooses heads and the other player chooses tails and the coin is tossed to see who wins.
The coin showed heads, so the visiting team kicked off.

heads   tails

hectare *noun*
A hectare is a unit of square measure or **area**, in the **metric system**. It is equal to 10,000 square metres or 2.471 **acres**.
The builder bought two hectares of land for his building site.

height *noun*
The height of an object is the distance from the bottom to the top.
The height of the cone is four and a half centimetres.

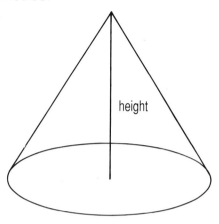

hemisphere *noun*
A hemisphere is a solid shape. A **sphere** cut in half through its centre results in two halves called hemispheres.
The dome of the church was shaped like a hemisphere.

hexagon *noun*
A hexagon is a flat, or plane, **shape** which has six straight sides and six angles. If all the sides and angles are equal, it is called a regular hexagon.
The interior angles of a hexagon are all equal to 120°.
hexagonal *adjective*

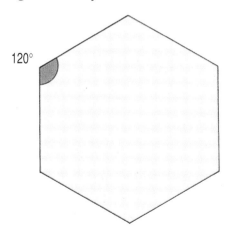

hexahedron ▶ shape

hexamine *noun*
A hexamine is a shape made up of six squares. Hexamines are often used as the basis of patterns.
He laid the paving stones in a hexamine shape.

highest common factor *noun*
The highest common factor, or HCF, of two numbers is the largest **factor** that divides into both of the numbers. The two numbers 12 and 18 can each be broken down into factors. 12 has the factors 1, 2, 3, 4, 6 and 12, and 18 has the factors 1, 2, 3, 6 and 18. The largest factor that is in both these sets is 6, so 6 is the highest common factor of 12 and 18.
The highest common factor of 8 and 6 is 2.

Hindu-Arabic numbers *noun*
Hindu-Arabic numbers originally came from the Hindus of India. Their number system of nine numerals was adopted by the Arabs. The Arabs introduced a tenth numeral, 'sifr', or zero.
Hindu-Arabic numerals are used throughout the Arabic-speaking world today.

hire *verb*
1. Hire means to pay money to use something. A customer can pay a car hire company to drive one of its cars. At the end of the hire period, the car must be returned to the company. The hirer does not become the owner of the car.
They hired a video camera for the weekend to film their new baby.
2. Hire also means to employ a person.
He hired five new staff last month to work in the café.

histogram ▶ bar chart

hoard *noun*
A hoard is an amount of money or valuables that has been saved.
The hoard of jewellery they found was worth £250,000.

Hong Kong dollar   *noun*
The Hong Kong dollar is the currency of Hong Kong. It is made up of 100 cents.

Hong Seng Index   *noun*
The Hong Seng Index is a number, or **index**, which indicates the movement of share prices on the Hong Kong Stock Exchange. The Hong Seng Index begins with a base number of 1,000. This increases or decreases in **proportion** to the movement of share prices. The Index can vary from minute to minute, depending upon the level of share activity.
The Hong Seng Index is studied to see how share prices in Hong Kong are moving.

horizontal   *adjective*
Horizontal means level. The opposite of horizontal is **vertical**.
The surface of a table is horizontal.

hour   *noun*
An hour is a unit of time. The number of hours measures how long something lasts. There are 24 hours in a day. Each hour is made up of 60 minutes.
The party lasted four hours.

hourglass   *noun*
An hourglass is a glass tube that measures time. The tube is very narrow at the middle and contains sand. It takes an hour for the sand to fall from the top half of the tube to the bottom.
Most people use a clock instead of an hourglass.

hyperbola   *noun*
A hyperbola is a kind of **curve**. It is the flat shape left behind when a cone is sliced vertically downwards. Because of this it is called a conic section. Circles and ellipses are also conic sections.
Some comets follow a hyperbola when they travel past the Sun.
hyperbolic   *adjective*

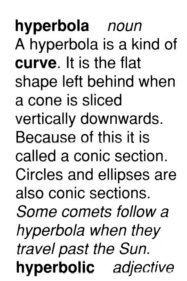
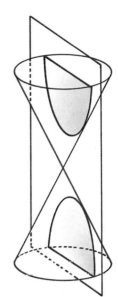

hyperinflation ► **inflation**

hypotenuse   *noun*
The hypotenuse is the side of a right-angled triangle that is opposite to the right angle. It is always the longest of the three sides. The hypotenuse is an important measurement when working on **Pythagoras' theorem**.
She measured the hypotenuse with her ruler.

hypothesis   *noun*
A hypothesis is a kind of guess. An investigation often begins with a hypothesis which is then tested. To test the hypothesis, an experiment involving all the facts must be carried out.
The hypothesis that iron is heavier than aluminium was found to be true.
hypothesize   *verb*

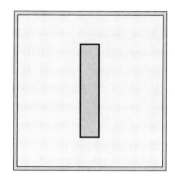

Icelandic króna (plural **krónur**)   *noun*
The Icelandic króna is the currency of
Iceland. There are 100 aurar to the króna.

icosahedron   *noun*
An icosahedron is a solid **shape**, or
polyhedron, that has 20 faces.
Each face of an icosahedron is a triangle.

IMF   *abbreviation*
The IMF is an international organization that
encourages international **trade**. Also, it can
help countries in difficulties with their
balance of payments. Countries that import
many goods may not have enough foreign
currency to pay for them all. The IMF will
sometimes lend money to solve this
problem. The letters IMF stand for
International Monetary Fund.
The country asked for a loan from the IMF.

imperial measures   *plural noun*
Imperial measures were the **standard**
measures of length, capacity and weight
formerly used in Great Britain. Many
countries around the world adopted these
measures. Today, most countries have
adopted the **metric system** of
measurement.
*The milk was delivered in the old imperial
measure of one pint.*

import   *verb*
To import means to buy **goods** or **services**
from suppliers in other countries. The
opposite of import is **export.**
*This year the company will import
1,000 tonnes of coffee from Brazil.*

imports   *plural noun*
Imports are **goods** or **services** bought from
suppliers in other countries. The opposite of
imports is **exports**. Often a country will try to
import as many goods as it exports.
*If imports increase, the balance of payments
will suffer.*

improper fraction   *noun*
An improper fraction is a fraction with a
larger **numerator** than **denominator**.
An improper fraction represents a **mixed
number** in fraction form. $\frac{13}{4}$, $\frac{26}{6}$, and $\frac{7}{3}$ are all
improper fractions.
*She turned the improper fraction into a
mixed number by dividing with the
denominator.*

inch (plural **inches**)   *noun*
An inch is a unit of length in the imperial
system of measurement. There are 12
inches in a **foot** and 36 inches in a **yard**. An
inch is equal to 2.54 centimetres in the
metric system.
*The compact disc measured just over
4 inches across.*

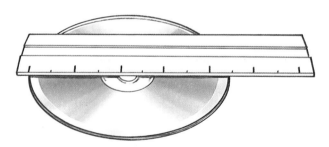

income   *noun*
Income is the money that a company or a
person receives. A person who has a job will
receive wages or a salary, and may also
earn **interest** on a bank account or a **gilt**.
A company will receive income from selling
goods or services to its customers. Another
word for income is revenue.
*His income during January was made up of
his salary and interest on investments*

income tax　*noun*
Income tax is a kind of tax. It is a tax paid by individuals but not by companies. The amount of income tax people have to pay depends on how much money they earn.
He paid £2,500 in income tax last year.

increase　*verb*
To increase is to get bigger. In mathematics, totals increase as numbers are added.
The amount of water in the pond increased after the rain.

index (plural **indices**)　*noun*
1. Index is another word for **power**.
In 2^3 the index of 2 is 3.
2. An index is a number used to measure changes in a country's **economy**. For example, the **Dow Jones Index** measures the change in prices on the New York Stock Exchange.
A Consumer Price Index measures the change in price of everyday goods and services.

Indian rupee　*noun*
The Indian rupee is the currency of India. The rupee is divided into 100 paise.

industry　*noun*
Industry is a general term used to describe manufacturing activity. Often the word refers to a group of companies making similar products, for example the steel industry or the motor industry.
The construction industry is in difficulty and few new houses are being built.
industrial　*adjective*

inequality　*noun*
Inequality describes two numbers of different value. Inequalities are like **equations**, but they use a 'greater than' sign (>) or a 'less than' sign (<) instead of an equal sign (=). The inequality 5 > 4 means that 5 is greater than 4.
The inequality 3 < 7 means 3 is less than 7.

infinity　*noun*
Infinity means going on for ever. It is represented by the symbol ∞. When recording a series of numbers, ∞ is often used to show they continue indefinitely.
The set of numbers can be written {0, 2, 4, 6, 8, 10, 12, ∞}.
infinite　*adjective*

inflation ► page 69

inflation ► page 69

ingot　*noun*
An ingot is a mass of metal in the form of a block or bar. It may be used at a later date to make a particular product. Gold, silver and other metals are formed into ingots.
They stored the gold as ingots before melting them down to make jewellery.

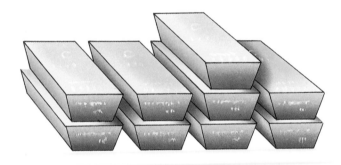

inheritance　*noun*
An inheritance is an amount of money or goods received as a gift on someone's death. Inheritances are often passed to children on the death of a parent.
When their parents died, their inheritance included a house.

insolvent ► **bankrupt**

inflation   *noun*

Inflation is a continual increase in prices. It is usually caused by an increase in the amount of money in **circulation**. There is more money but it can buy fewer goods and services. Hyper-inflation is very severe inflation. The opposite of inflation is **deflation**.

The rate of inflation is coming down gradually.

inflationary   *adjective*

All these things contribute to the price of bread.

If their cost goes up, the price of bread goes up.

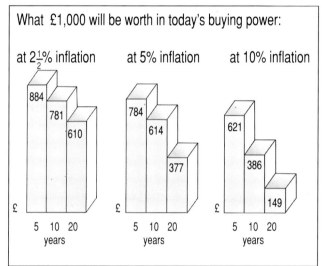

What £1,000 will be worth in today's buying power:

at $2\frac{1}{2}$% inflation | at 5% inflation | at 10% inflation

884 781 610
784 614 377
621 386 149

£   5 10 20 years
£   5 10 20 years
£   5 10 20 years

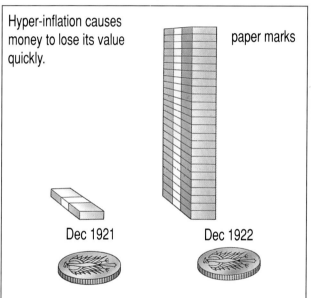

Hyper-inflation causes money to lose its value quickly.

paper marks

Dec 1921   Dec 1922

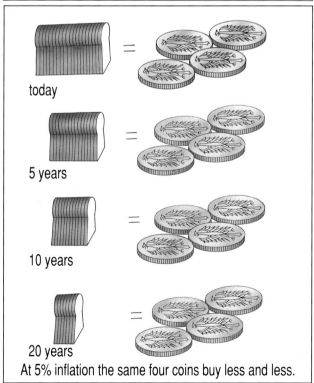

today

5 years

10 years

20 years

At 5% inflation the same four coins buy less and less.

In the space of a year, post-war German paper money became worth a fraction of what it had been worth. It required a basket of money to buy a loaf of bread.

insurance *noun*
Insurance is a payment of money made so that if there is a misfortune in the future, a sum of money is paid as **compensation**. A house owner might insure a house. A **premium** is paid in case the house is damaged.
The insurance on his car cost £250 per year.
insure *verb*

integer *noun*
An integer is a positive or a negative number. For example, 2, 5, 9, –3, –7 and 0 are all integers. Fractions, or numbers with fractions, are not integers. Another term for integer is whole number.
The sum of two integers is always another integer.

intercept *noun*
The intercept of a **graph** is the place where the line on the graph crosses one of the **axes**. The graph of the **linear equation** $y = x – 2$ crosses the x-axis at $x = 2$.
The intercept with the y-axis is at $y = –2$.

interest *noun*
Interest is the price the **lender** charges when someone borrows money. It is a percentage of the total loan, or amount borrowed. Interest can be calculated as **simple interest** or **compound interest**. It is paid at regular intervals, perhaps each month or each year.
He had to pay interest on his loan.

interior angle *noun*
An interior angle is an angle on the inside of a shape. The interior angles of a rectangle are all right angles.
The interior angles of a triangle always add up to 180°.

international *adjective*
International describes things that are made up from or affect more than one country.
The United Nations is an international organization.

International Monetary Fund ► IMF

interpolation *noun*
Interpolation is a way of working out numbers that are between other numbers. Interpolation can be used to find the price of 25 nails from a price list showing the price of nails in groups of 10.
The price of 25 nails was 24 pence.

Number of nails	Price in pence
10	10
20	19
30	27
40	34

intersection *noun*
1. An intersection is the place where two or more lines cross.
The teacher placed a cross at the intersection of the lines.
2. The intersection of two **sets** is the collection of all the things that belong to both sets. If one set contains pencils and the other set contains green things, the intersection of the two sets is the collection of all the green pencils. The intersection of two sets can be worked out on a **Venn diagram**.
When the objects were sorted, some of them overlapped in the intersection.

Venn diagram

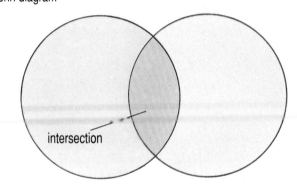

intersection

inventory *noun*
An inventory is a list of **goods** or possessions.
The shop made an inventory of its stock every six months.

inverse proportion   *noun*
Inverse proportion means that when one thing increases, something else decreases. For example, if three gardeners take four hours to dig a garden, six gardeners will take only two hours.
The number of gardeners and the time they need to do a job are in inverse proportion.

invert   *verb*
To invert is to turn something upside down or back to front. To invert a fraction, the **numerator** and **denominator** change position. Shapes can also be inverted.
If you invert a triangle, it is still a triangle.

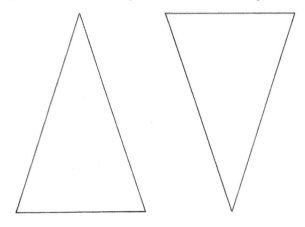

investment   *noun*
1. An investment is something which a business spends money on in order to earn more money. For example, a business might invest in a new machine so that it can make its products more cheaply and efficiently.
The company plans an investment of about two million pounds in new machinery next year.
2. An investment is a financial **asset** bought by an individual or a company in order to earn **income** from it or to make a **capital gain** on selling it. For example, a person might buy **shares** in a large company. **Dividends** would be paid out from the company at regular intervals.
Her investment in shares earned her an income of £230 last year.
invest   *verb*

investment trust   *noun*
An investment trust is a company which uses its money to buy a wide range of **shares** and other **investments**. It will purchase from many different sources to reduce its **risk**. By choosing investments wisely, the investment trust can earn a good **income** as well as increase its **capital**. Individuals can buy shares in an investment trust. They are then entitled to a share of the income it produces.
He paid £2,000 to buy shares in an investment trust.

invoice ► **bill**

Iraqi dinar   *noun*
The Iraqi dinar is the currency of Iraq.
An Iraqi dinar is made up of 1,000 fils.

Irish punt   *noun*
The Irish punt is the currency of Ireland.
There are 100 pence in a punt.

irregular shape   *noun*
An irregular shape is one in which all the sides are not of equal length, and interior angles are of different sizes.
The boundary of the city formed an irregular shape.

isosceles triangle
noun
An isosceles triangle is a triangle that has two sides the same length. Two of the angles of an isosceles triangle are equal. The name isosceles comes from the Greek word 'isos' which means equal.
The tower was designed as a steep sided isosceles triangle.

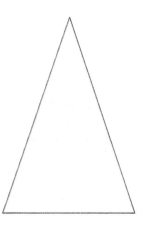

issue *noun*
An issue is an amount of **shares** or other **investments** which a company makes available for purchase.
Two thousand shares were made available by the company in the last share issue.

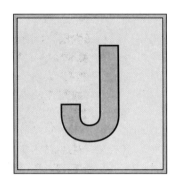

job *noun*
A job is something that has to be done. Some jobs are done to earn money, but people often have to do jobs without getting paid.
A person who has a job is someone who is employed.

joint account *noun*
A joint account is a bank or building society **account** created for the use of two or more customers. Each person named on the account can make **deposits** or **withdrawals**.
The husband and wife opened a joint account with the local bank.

joint venture *noun*
A joint venture is a business or company that is set up by two or more individual companies. A joint venture is often formed when a single company does not have enough money or resources, or is prevented for some other reason from setting up in a market on its own. It joins forces with a second company to achieve this. Both companies hold **stock** and make **profits** from the joint venture company.
The company set up a joint venture with a local company to make cement in Dubai.

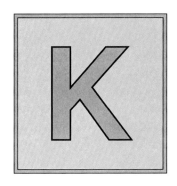

Kenyan shilling *noun*

The Kenyan shilling is the currency of Kenya. One shilling is divided into 100 cents.

kilo- *prefix*

Kilo- is a prefix that means a thousand. It is usually shortened to k. A kilogram is a thousand grams and is written in its short form as kg. A kilometre, or km, is a thousand metres.

The bag of pears weighed two kilograms.

kilogram *noun*

A kilogram is a measure of **weight** and mass in the **metric system**. It is made up of 1,000 grams.

They ate a kilogram of strawberries.

kilometre *noun*

A kilometre is a **metric** measure of length. It is made up of 1,000 metres.

The distance across the islands was 7 kilometres.

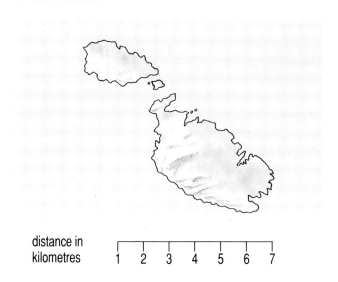

distance in
kilometres   1 2 3 4 5 6 7

koruna *noun*

The koruna is the currency of the Czech and Slovak Republics. It is made up of 100 haléru.

krona *noun*

The krona is the unit of currency of several countries, including Iceland and Sweden. In Sweden, the krona is made up of 100 öre. In Iceland, there are 100 aurar to one krona.

krone *noun*

The krone is the unit of currency of several countries, including Norway and Denmark. In both countries, 100 ære make one krone.

Krugerand ► rand

kyat *noun*

The kyat is the currency of Burma. There are 100 pyas in a kyat.

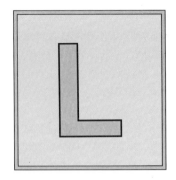

landlord *noun*
A landlord is a person or company who owns a building and allows someone else to live in it or do business in it. In return for this, the person who occupies the building pays **rent** to the landlord.
The landlord charged a rent of £150 per month.

language ► page 76

latitude *noun*
Latitude is a measurement of how far north or south of the Equator a place is on the Earth's surface. It is measured in **degrees** north or degrees south. A place on the Equator has latitude 0°. The North Pole has latitude 90° north. Lines of latitude are imaginary lines which run from the North to the South pole.
The latitude of Philadelphia in the United States is 40° north.

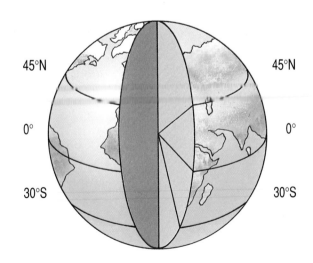

lattice ► grid

leap year *noun*
A leap year is a year with 366 days instead of 365 days. The extra day is added to February. Every four years, February has 29 days instead of 28.
If the last two digits of a year are divisible by four, the year is a leap year.

lease *noun*
A lease is a **contract** by which the owner of an **asset** allows another person to use it in exchange for regular payments. The lease outlines what **rent** will be paid, and the length of time the agreement will last.
The company entered into a lease agreement to obtain the use of a machine for monthly payments of £270.
lease *verb*

ledger *noun*
A ledger is a collection of **accounts**. It is a financial record of everything a business has bought or sold. The ledger may be written down in a book, or it may be held on a computer.
The sales ledger contains a list of the company's customers who buy goods on credit.

legal tender *noun*
Legal tender means coins and notes that may be used to buy things.
In Britain, the farthing used to be legal tender, but it is no longer in use.

lek *noun*
The lek is the currency of Albania. One lek is made up of 100 qintars.

lend *verb*
To lend means to allow someone else to borrow money or goods which must be returned later. In business, lending usually means handing money to someone on condition that it is repaid in the future.
Until the **loan** is repaid, it is normal to pay **interest** on it.
The bank will lend the company £10,000.

length ▶ distance

less *adjective*
Less means not as many as, or fewer. It can also mean subtract. The mathematical **symbol** stands for less than. The − symbol stands for subtract, or minus.
Five is less than nine and can be written 5 < 9.

letter of credit *noun*
A letter of credit is a **guarantee**, or legal promise. It authorizes the named person or company to receive a certain amount of cash or **credit**.
Letters of credit are sometimes used to guarantee payments for goods to be exported.

leu (plural **lei**) *noun*
The leu is the currency of Romania. There are 100 bani to one leu.

lev (plural **leva**) *noun*
The lev is the currency of Bulgaria. The lev is divided into 100 stotinki.

liability *noun*
Liability is an amount of money that an individual or a business owes to another. If a business buys goods on **credit**, it has a liability to the company that supplied the goods. If an individual borrows money from a bank, they have a liability to that bank. The liabilities of a business are listed on its **balance sheet**.
The company decided to reduce its liabilities by repaying the money it owed to the bank.

limited company *noun*
A limited company is a **company** whose **shareholders** have no responsibility for its liabilities, or debts, once their shares are fully paid up.
The limited company went into liquidation, but the shareholders had no liability since their shares were fully paid.

line *noun*
A line is the path traced by a moving point. It has length but no width or thickness.
A pencil was used to draw a line between the two points on the map.

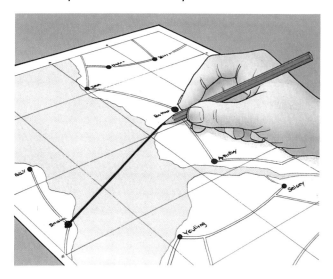

linear *adjective*
Linear describes a straight line. The linear distance between two places is the distance between them measured in a straight line. An equation is linear if it makes a straight line when it is drawn on a graph.
A linear measurement can be used to find the distance between two towns.

linear equation *noun*
A linear equation is a kind of equation used in **algebra**. An equation like $y = 2x + 3$ is a linear equation because when it is drawn on a **graph** with y on the vertical **axis**, and x on the horizontal axis, it makes a straight line.
Linear equations are the simplest equations used in algebra.

liquidate *verb*
To liquidate means to sell off the **assets** of a company to raise cash. This may happen when the company is **bankrupt** and cannot pay its **debts**. The money raised from selling off the assets is used to pay as much of the company's debts as possible.
The directors decided to liquidate the company.

language *noun*

The language of number refers to the names given to numbers and mathematical terms. There are around 3,000 languages in the world, and each one has its own set of words to refer to numbers.

Languages that have developed from shared origins sometimes use similar words for numbers.

The googol

Googol is the name given to a one followed by one hundred zeros. Written in full, a googol looks like 10000000000000000000000000000000000000000000000000000000000000000000000000000000000000000000000000000000000000000000000000. It can be written more economically as 10^{100}. The term googolplex is used to mean ten multiplied by itself a googol times. Mathematicians find it useful to have words to refer to the huge numbers they sometimes work with.

English	Chinese	French	German	Arabic	Hindi
One	一	Un	Eins	واحِد	एक
Two	二	Deux	Zwei	إِثْنان	दो
Three	三	Trois	Drei	ثلاثَة	तीन
Four	四	Quatre	Vier	أَرْبَعة	चार
Five	五	Cinq	Fünf	خَمْسَة	पाँच
Six	六	Six	Sechs	سِتَّة	छः
Seven	七	Sept	Sieben	سَبْعة	सात
Eight	八	Huit	Acht	ثَمانِية	आठ
Nine	九	Neuf	Neun	تِسْعَة	नौ
Ten	十	Dix	Zehn	عَشَرَة	दस

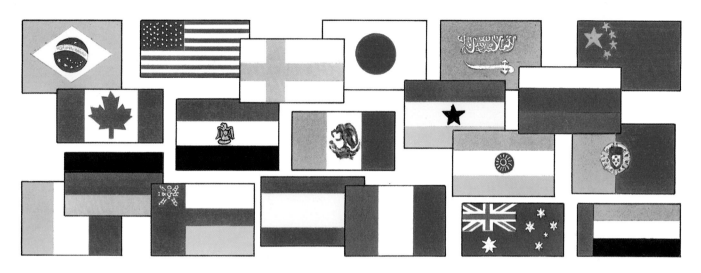

Spanish	Russian	Italian	Portugese	Ashanti
Uno	Один	Uno	Um	Eko
Dos	два	Due	Dois	Eno
Tres	три	Tre	Três	Esa
Cuatro	четыре	Quattro	Quatro	Enae
Cinco	пять	Cinque	Cinco	Innum
Seis	шесть	Sei	Seis	Insia
Siete	семь	Sette	Sete	Nso
Ocho	восемь	Otto	Oito	Inwotwi
Nueve	девять	Nove	Nove	Enkoro
Diez	десять	Dieci	Dez	Edu

lira (plural lire) *noun*

The lira is the unit of currency of several countries, including Italy and Turkey. The Turkish lira is equal to 100 kurus or piastres.

listed company *noun*

A listed company is a company which has obtained permission to have its shares included in the daily official list of a stock exchange. Stock in the company may be freely bought and sold.
He invested in a listed company.

litre *noun*

A litre is a metric unit of capacity. The number of litres measures how much space a liquid takes up.
He bought 20 litres of petrol.

loan *noun*

A loan is an amount of money that someone borrows. The money has to be paid back. When a bank lends money, it usually charges **interest** on the loan.
He took out a loan for $10,000 to buy a car.

locus (plural loci) *noun*

A locus is the path of a point, line or surface.
The locus of a cannonball is a parabola.

logarithm *noun*

A logarithm is a number used in repeated multiplication. It is known as a **power**. To find the logarithm of a number its **base** must be known. The logarithm of 1000 in base 10 is 3 because 10 to the power of 3, or 10^3, = $10 \times 10 \times 10 = 1,000$.
Can you work out the logarithm 2^4?

logic *noun*

Logic is a way to think in order to solve problems. All the facts are first gathered together and then built on to work out new facts. For example, in the diagram below, it is possible to criss-cross from one island to another using each bridge only once.
Some problems in mathematics need logic to work them out.

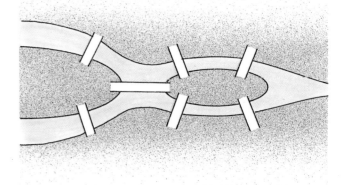

logo *noun*

A logo is a symbol which a company uses to distinguish itself from other companies. It is a form of **trademark**.
The airline company used a red wing as its logo.

long division *noun*

Long division is a way of setting out a division **sum** when the divider is greater than 10. The full number is divided in parts, the sum being worked from the high value to the lower value numbers.
Long division is used to work out the sum $289 \div 13$.

long multiplication *noun*

Long multiplication is the method used to multiply by a number greater than 10. Instead of multiplying by the whole **multiplier** at once, it is broken down into smaller value numbers that are easier to use. For example, to multiply by 256, the sum of three multiplication sums is needed – by 6, by 50 and then by 200.
To multiply a number by 451, a long multiplication sum is used.

longitude   *noun*

Longitude is a measurement of how far east or west of Greenwich a place is on the Earth's surface. Greenwich is a place near London, in the United Kingdom. Longitude is measured in degrees west or east of Greenwich. Greenwich has longitude 0°.
The Greenwich meridian is represented by a brass line on the ground at Greenwich.

loss   *noun*

A loss occurs when the **expenditure** of a business is greater than its **income**. The opposite of loss is **profit**.
The company made a loss last year because sales were low.

lottery ► page 80

lowest common multiple   *noun*

The lowest common multiple, or LCM, of two **digits** is the smallest number that they both divide into without a **remainder**. The two numbers 4 and 6 both divide into 12, so 12 is their lowest common multiple.
The lowest common multiple of 6 and 4 is 12.

Luxembourg franc   *noun*

The Luxembourg franc is the currency of Luxembourg. One franc is made up of 100 centimes.

luxury   *adjective*

Luxury is used to describe goods or items which are enjoyable to own but which are not necessary to have in order to live. An expensive car might be considered a luxury item where public transport can be used.
They spent too much money on luxury clothes and cars.

Lydian coin   *noun*

Lydian coins were the first coins ever made. Lydia was an area in the country now called Turkey. The coins were bean-shaped lumps of a metal called electrum. This was a mixture of gold and silver. The coins were stamped with the king's head to **guarantee** their value. This idea was soon adopted by traders from other countries and the idea spread.
Lydian coins were circulated as long ago as 545 BC.

lottery *noun*

A lottery is a way of raising money by selling numbered tickets. Numbers are drawn at **random**, and holders of the winning tickets receive a prize. Any game in which the winner is picked at random is a kind of lottery.

He bought ten tickets in the lottery, but did not win anything.

The winning ticket must be presented to collect the prize.

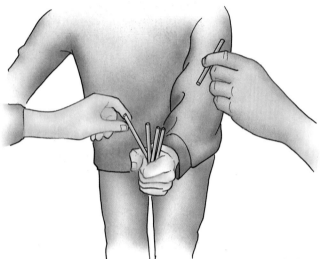

The person who picks the short straw is usually the unlucky one who has to do something. This is called choosing lots.

Numbers are mixed up in a barrel, and winning numbers picked at random.

Some lottery tickets let you know immediately whether you have won.

Lottery tickets are sometimes sold on the streets.

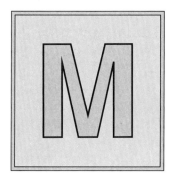

magic square ► page 82

magnetic strip   *noun*
A magnetic strip is a strip of magnetized metal found on the back of a credit or bank card. It holds the **code** which is used to transfer information about the card owner to a machine such as a cash dispenser or counter till. A magnetic strip identifies the card owner.
The magnetic strip on the cash card had worn out, so the card was not accepted by the cash dispenser.

magnitude   *noun*
Magnitude means size. It is used when referring to measurements.
The company saw huge profits owing to the magnitude of the order.

mail order   *noun*
Mail order is a way of buying **goods** without having to go to a shop. A mail order company advertises its goods in a catalogue. A person who wants to buy the goods either writes to the company, or telephones it. A person pays for the goods by mailing a **money order** or **cheque**, or by **credit card**. The company then supplies the goods by post.
She bought a new coat by mail order.

manufacturer   *noun*
A manufacturer is a company that makes goods. The manufacturer will usually sell these goods to a shop, which will sell them to customers.
The manufacturer made sewing machines.

map   *noun*
A map is a drawing that represents part of the Earth's surface. There are many kinds of map. Some maps show whole countries, and some show details of roads in a town or smaller area. Maps of the sea are called charts. These show how deep the sea is, so that ships can sail on a safe course.
The hikers used a map to find their way across country.

mapping   *noun*
Mapping is a diagram which shows the connection between two sets of numbers. It is also called a **function**.
They used mapping to link the individual numbers of the set together.

margin   *noun*
A margin is a form of **profit**. For example, a toyshop must buy the toys it sells. The margin for each toy is the difference between the price the customer pays and the total costs the shop must pay in order to **stock** it.
The margin on the toy was £2.

mark ► **deutschmark**

market *(noun)* ► page 84

market   *verb*
To market is sell.
The clothes company will market their new coats in April.

market research   *noun*
Market research means finding out what customers want. This is usually done by asking individuals to reply to questions in a **survey** about what they buy. A supermarket may ask customers what kind of soup they would prefer to buy. The supermarket can then stock the kind that they know customers like.
The supermarket's market research showed that most customers preferred to buy tomato soup.

magic square   *noun*

A magic square is a kind of **matrix**. In a magic square, each row and each column adds up to the same number. Adding the numbers diagonally from corner to corner also gives the same number. A magic square can be any size.

In a 4 by 4 magic square, there are 880 different ways of arranging the numbers 1 to 16.

Rows in a 3 by 3 magic square always add up to 15. Magic squares like this were known 2,000 years ago in China.

16	3	2	13
5	10	11	8
9	6	7	12
4	15	14	1

This magic square was devised by the great German artist, Albrecht Dürer. The rows all add up to 34, and the numbers in the pink squares at the bottom indicate the year Dürer made the square, 1514.

marketing   *verb*
Marketing means making buyers aware that a product is available. It is usually done by people involved with sales. Marketing may include making a brochure or creating an eye-catching display.
The company is marketing its products using an attractive display.

markka   *noun*
The markka is the currency of Finland. It is made up of 100 pennies.

mathematics   *noun*
Mathematics is the science of numbers. It is the study of patterns, measurements and relationships between numbers. Branches of mathematics include **algebra**, **arithmetic**, **calculus**, **geometry**, **logic**, **statistics** and **trigonometry**.
Mathematics is a complex science.
mathematical   *adjective*

matrix *(plural* **matrices***)*   *noun*
A matrix is a set of numbers arranged in a rectangle. It is set out as a numerical **table** and is used to organize large amounts of information in order to make them easier to read and understand. The output from a **spreadsheet** is a kind of matrix.
They used a matrix to show how many and what kinds of cars were sold in the country.

maximum   *noun*
The maximum of a set of numbers is the largest number in it. The maximum of the set {2, 6, 4, 8, 1, 3} is 8.
The maximum number of people a lift can carry is 12.

mean ► **average**

measure   *verb*
To measure is to find out the size or extent of something. Lengths can be measured in **metres**, and weights can be measured in **grams**. Other examples of measures are **Celsius** which is used to measure temperature, and ampere which is used for measuring the flow of electric current.
Angles are measured with a protractor in degrees.

measurement ► page 86

median   *noun*
1. A median is the middle of something. In a set of numbers that are sorted in order of size, the median of the numbers is the middle one in the sequence. The median of {2, 4, 6, 7, 9, 12, 25} is 7.
They took the July temperature in Morocco as a median guide.
2. The median is a line drawn from the **vertex**, or **apex**, of a triangle to the mid-point of the opposite side.
The median divided the base of the scalene triangle.

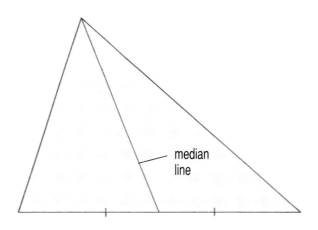

median line

market *noun*

1. A market is a place where people buy and sell things.

She sold eggs from a stall in the market.

2. The market can also describe a group of buyers who are likely to be most interested in buying a particular product or range of products. For example, a market might refer to the group of young people under 21 years of age.

The designer presented new ideas for the children's toy market.

A street seller sells goods from his bicycle.

The shelves are often well stocked in a supermarket.

There are many different stalls in a souk.

Merchants sell goods from their boats in a floating market.

A roadside stall sells medicines.

Stocks and shares, not goods, are bought and sold at a stock market.

measurement _noun_

A measurement is a way of finding out the size, weight or extent of something. Measurements are made according to a scale of **units**. The base units in the SI **metric system** of measurement are the metre to measure length, the kilogram for weight or mass, the second for time and the kelvin for temperature.

He made a careful measurement of the doorframe to make sure that he had enough wood to fix it.

Size
Length, area and volume are measured in metres.

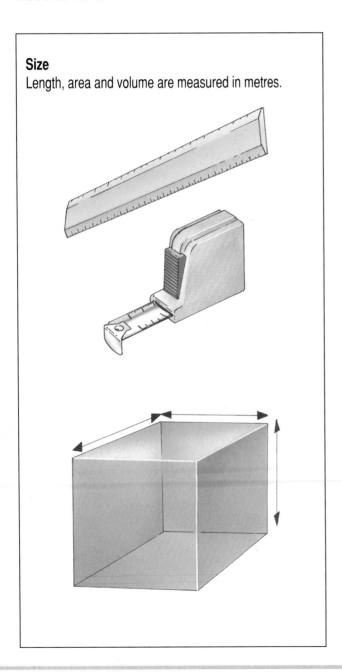

Volume
The volume of liquid is measured in litres.

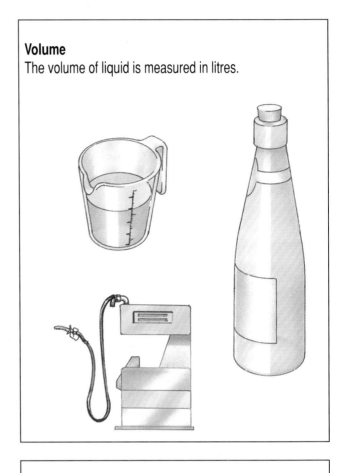

Temperature
Thermometers measure temperature on the Kelvin, Celsius or Fahrenheit scale. The Celsius scale is based on the Kelvin scale.

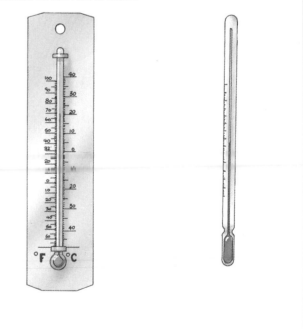

Weight or mass

A metal cylinder weighing exactly one kilogram is the standard for all units of weight and mass. The cylinder is kept in France.

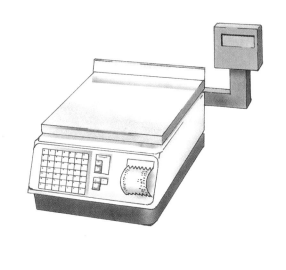

Other pieces of equipment make different measurements

watch measures time

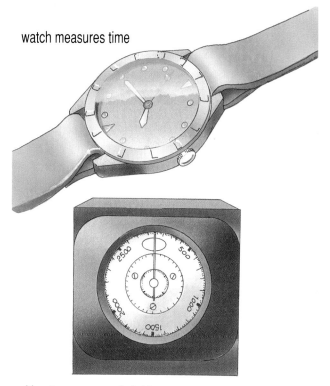

altimeter measures height

speedometer measures speed

barometer measures air pressure

member *noun*
1. A member is an object or value that belongs to a **set**. In a set of cups, each cup is a member of the set of cups.
The parachute club had 24 members.
2. In business, a member of a company is one of the people who own the company. Another word for member is **shareholder**.
The members of the company voted to pay a dividend on the profit.

merchandise ► goods

merchant ► trader

merchant bank *noun*
A merchant bank is a bank that does business with companies, not individuals. If a **public company** wants to borrow money, it usually goes to a merchant bank.
The company asked for advice from its merchant bank.

merger *noun*
A merger takes place when two or more companies join together to form a single company. Two companies that carry on the same kind of business might join forces to try to get a bigger share of the **market**.
The two companies expect to make bigger profits with the merger.

metre *noun*
A metre is a metric unit of length. It is used to measure how long, wide or high something is. There are 100 centimetres in a metre.
The hedge was two metres high.

metric system *noun*
The metric system is a set of units used for **measurement**. The three main units are the **gram**, which measures weight, the metre, which measures length, and the **litre** which measures capacity. Each unit increases by tens and by tens of tens.
The metric system is used in many countries of the world.

Mexican peso *noun*
The Mexican peso is the currency of Mexico.

mile *noun*
A mile is an **imperial measure** of length. A mile measures 1,760 yards and is equal to 1.61 kilometres in the **metric system**.
He walked several miles before becoming tired.

milli- *prefix*
Milli- is a prefix that means one thousandth. A millimetre is a thousandth of a metre, so there are 1,000 millimetres in a metre.
The bottle contained only 16 millilitres of water.

milligram *noun*
A milligram is a measure of weight in the **metric system**. There are 1,000 milligrams in a **gram**.
The gold dust was carefully measured in milligrams.

millilitre *noun*
A millilitre is a measure of volume in the **metric system**. There are 1,000 millilitres in a **litre**.
There were 200 millilitres of liquid in the test tube.

millimetre *noun*
A millimetre is a measure of length in the **metric system**. There are 1,000 millimetres in a metre.
The insect measured only a few millimetres in length.

milling *noun*

Milling is a pattern of small grooves on the edges of some coins. When coins were made of gold, people were able to scrape a little of the gold off the edge of the coin. . Milling was invented so that any scraping could be easily seen.

Although most modern coins are not made of gold, many still have milling.

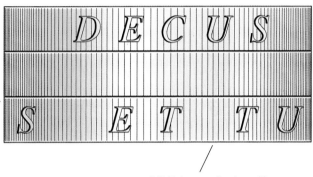

3 British pound coins with milled edges

million *noun*

A million is a thousand thousands. It is written 1,000,000.

The country's population in 1985 was 75 million.

millionaire *noun*

A millionaire is a person who has a million units of **currency**. In the United States of America, a millionaire has a million dollars.

The British millionaire had a million pounds.

minimum *noun*

The minimum is the smallest, or least, amount. The minimum of a **set** of numbers is the smallest number in the set. The minimum of the set {2, 6, 4, 8, 1, 3} is 1.

The minimum temperature today was 5°C.

mint ► page 90

minus sign *noun*

Minus is the sign − used to indicate that the following number is to be **subtracted** or that it is a **negative number**.

Temperatures with a minus sign are cold.

minute *noun*

1. A minute is a unit of time. There are 60 minutes in an hour. Clock-makers divided the hour into 60 minutes when clocks became accurate enough to need smaller units than the **hour**.

It took 10 minutes to walk to school.

2. A minute is a unit of an angle. It is one sixtieth of a **degree**.

A circle is divided into 360 degrees and each degree is divided into 60 minutes.

miser *noun*

A miser is a person who saves money and will not spend it. The opposite of a miser is a spendthrift.

The miser would not give any money to charity.

mixed number *noun*

A mixed number is a **whole number** and a **fraction** together. $3\frac{3}{4}$ and $4\frac{5}{9}$ are both mixed numbers.

He changed an improper fraction into a mixed number.

mobius strip *noun*

A mobius strip is a strip of material which is given a one-half twist and joined at the ends. It has only one side and one edge, and stays in one piece when it is split down the middle.

A mobius strip has interesting geometric properties.

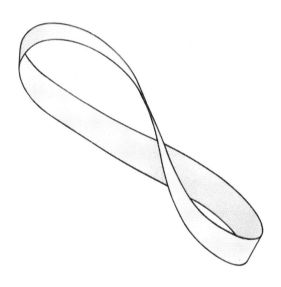

mint *noun*

A mint is a place where coins are made. The government carefully controls the making of coins. The mint is also used to store coins before they go into circulation. When coins are minted, they are counted into bags on a 'telling' machine. A label is attached to show how many coins there are and of what **denomination**. They are then taken to a stronghold ready for despatch.

They made pound coins at the mint.

An artist makes a large model of the coin in plaster.

This electrotype copy is put on a reducing machine. It is scanned in a spiral by a tracer. The tracer movements are conducted by a bar to a rotating cutter. The cutter copies each movement, at the size of the finished coin. It is cutting into a block of steel. This steel copy is the master punch, with the features of the coin in relief. The master punch is used to make working punches, which are used to make working dies. The details are gone over to get rid of any flaws and the working dies are polished.

A mould of the model is placed in an electroplating bath. Nickel and then copper are deposited on it.

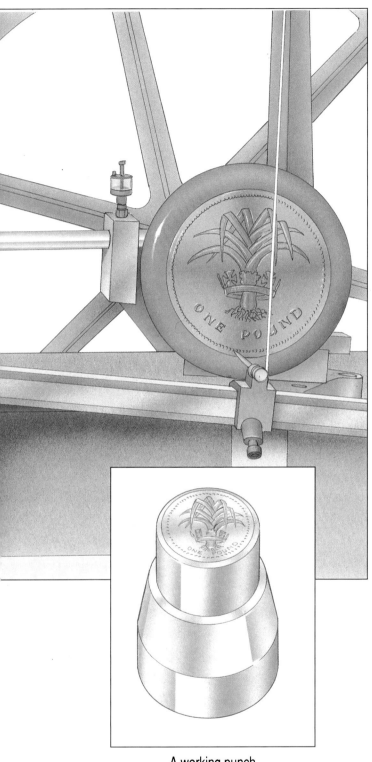

A working punch.

Some of the coins need to be examined by hand.

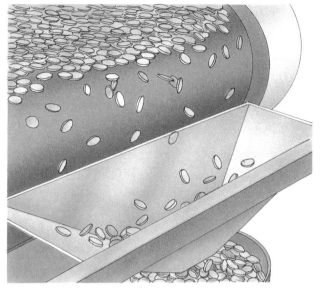

Blank coins are softened by passing them through a furnace.

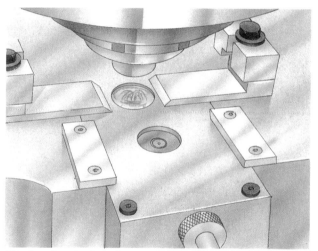

The head and tail design are stamped onto the blank at the same time. A collar holds the blank in place.

mode *noun*
The mode of a **set** of numbers is the number that is most frequent. The mode of the set {3, 6, 4, 8, 6, 4, 3, 4, 9} is 4 because there are more 4's than any other number.
To find out the most common shoe size, the shop manager worked out the mode of all the sizes.

model ► page 93

modulus *noun*
The modulus is the absolute value of a number, or its value regardless of any **signs** it may have around it. The modulus of a **positive number** is the number itself, so the modulus of 4 is 4. The modulus of a **negative number** is the number without the **minus** sign, so the modulus of −5 is 5. The modulus is always a positive number.
They worked out the modulus of −6 to be 6.

monetary policy *noun*
Monetary policy describes the actions taken by a government to influence the way a country's **economy** operates. It usually involves the control of the money supply, and of **credit** and **interest** rates. Monetary policy is used to help the economy grow, to keep as many people employed as possible and to keep prices and wages at the correct level.
The government changed its monetary policy in order to help the country out of recession.

money *noun*
Money means any coins, notes, cheques or other things that can be used to buy goods or services. Beads, cocoa beans, shells and stones are some of the things that have been used as money in the past. Silver and gold were used more widely. They were the main forms of money because they were convenient and hardwearing.
He does not have enough money to go on holiday.
monetary *adjective*

money order *noun*
A money order is a form which promises to pay a certain amount of money on demand. The money is paid to the person whose name is written on the order. Money orders can be bought from banks, post offices and some other businesses. They are a quick way of transferring small amounts of money.
Their grandmother bought them money orders for their birthdays.

monopoly *noun*
A monopoly is formed when a product or service is only available from one company. A monopoly often allows a company to charge high prices for its products.
The opposite of monopoly is competition.

month *noun*
A month is a unit of time. The year is divided into 12 months, although not all the months are of the same length.
A lunar month is the time it takes for the moon to circle the Earth.

mortgage *noun*
A mortgage is an amount of money given as a **loan** to buy a building or land. It is usually provided by a bank or **building society**. Mortgage repayments are made up of the money borrowed plus **interest** and are repaid every month for many years.
Their mortgage was repaid over 25 years.

model   *noun*

A model is a mathematical projection or design. Combinations of numbers or a **formula** can also be used as models. A scale model is an exact copy of something reproduced in a smaller or larger size. The scale is usually given as a **ratio**. Computers are often used to build models of things that change or move, because of the high number of mathematical calculations needed to describe them.

Scientists do not know what an atom really looks like, so they use models to explain how atoms behave.

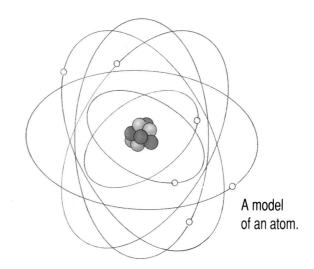

A model of an atom.

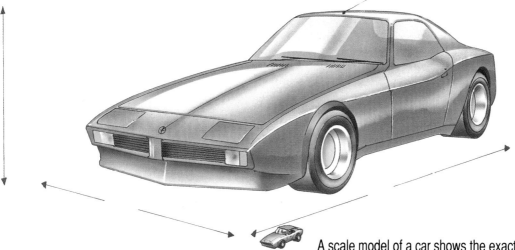

A scale model of a car shows the exact dimensions of the original. It is projected mathematically.

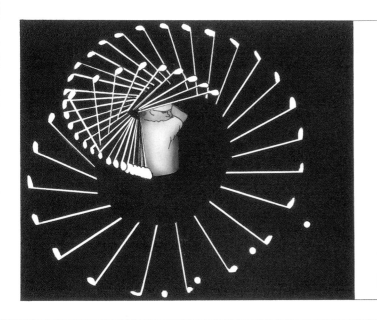

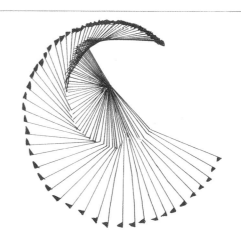

This computer model shows the exact movement of the golfer using a graphic image.

multinational *adjective*
Multinational describes a company that has branches in more than one country.
The multinational company opened a new branch in Peru.

multiple *noun*
The multiple of a number is any whole number into which it will divide exactly.
He discovered the first 5 multiples of 3 were 6, 9, 12, 15, 18.

multiplicand *noun*
The multiplicand is the number that is being multiplied by the **multiplier**. It is the first number in the sum.
In the sum 426 x 8, 426 is the multiplicand.

multiplication *noun*
Multiplication is one branch of arithmetic. The **sign** for multiplication is X. Multiplication is a short way of adding groups of numbers together. To find the total number of biscuits in four packets of five biscuits, the equation $4 \times 5 = 20$ is shorter than $5 + 5 + 5 + 5 = 20$.
They used multiplication to work out how many eggs were in five boxes of one dozen each.
multiply *verb*

multiplier *noun*
The multiplier is the number by which another number is multiplied. It is the second number in the sum.
In 426 x 8, eight is the multiplier.

naira *noun*
The naira is the currency of Nigeria. There are 100 kobo in a naira.

nano- *prefix*
Nano- means one thousand millionth. It is an extremely small measurement. Astronomers make some of their measurements in nanoseconds.
The prefix nano- is used in the metric system.

Napier's bones *noun*
Napier's bones is a method of working out **long multiplication**. The **formula** is named after a Scottish mathematician, John Napier, who lived from 1550 to 1617. Napier first designed his system on bones, but today 10 wood or board strips are used. Nine carry the multiplication tables. One is an index.
The strips are matched side by side to work out an answer.

national bank *noun*
A national bank is a bank that operates under the special control of a country's government. Some national banks are owned by their governments.
The Bank of England is a national bank.

national debt *noun*
The national debt is the total amount of money the government of a country owes. The level of the debt depends on how much the government has borrowed, and on how much it has paid back.
The country's national debt increased as the government borrowed more money to pay for imports.

natural number   *noun*
A natural number is a number that is greater than zero. 1, 2, 3, 4, 5, and 6 are natural numbers. Other names for natural numbers are **whole numbers** or **positive numbers**.
Natural numbers are always positive numbers.

negative number   *noun*
A negative number is a number that is less than zero. Adding a negative number to a **positive number** always reduces the value of the positive number. For example, –3 is a negative number. If –3 is added to 5, the answer is worked out as –3 + 5 = 2.
The sum of two negative numbers is always negative.

negotiate   *verb*
To negotiate is to discuss and form an agreement over a possible **deal**. In business, a sales person might negotiate the price of certain goods with a **buyer**.
They sat round a table ready to negotiate the terms of the contract.
negotiation   *noun*

net (noun) ► page 96

net   *adjective*
Net describes the value of something after any necessary **deductions** have been made. If a product has a price of 60 pesos, but the seller offers a **discount** of 5 pesos, then the net price is 55 pesos. If someone's wages are $200 per week, but the employer deducts $40 for tax, the net pay is $160.
After deducting taxes, the company worked out its net profit.

net profit   *noun*
Net profit is the amount of profit earned by a business after all **expenditure** has been deducted. It will always be less than the **gross profit**.
The gross profit earned by the company was £35,000, but after costs had been deducted the net profit was only £10,000.

new sol   *noun*
The new sol is the currency of Peru.

New Zealand dollar   *noun*
The New Zealand dollar is the currency of New Zealand. It is made up of 100 cents.

nickel   *noun*
A nickel is a coin used in the United States of America and Canada. It is worth five cents.

Nikkei Average Price Index   *noun*
The Nikkei Average Price Index is an **index**, or number used to indicate the movement of share prices on the Tokyo Nikkei Stock Exchange. It uses the prices of the top 100 shares and changes minute by minute as share dealing takes place.
The Nikkei Average Price Index shot up as investors bought more and more shares.

Norwegian krone   *noun*
The Norwegian krone is the currency of Norway. It is made up of 100 öre.

note ► bank note

number   *noun*
A number is a quantity. The number of sweets in a bag can be worked out by counting them. The first 10 numbers are 1, 2, 3, 4, 5, 6, 7, 8, 9, 10.
The number of wheels on a car is 4.
numerical   *adjective*

numeral ► page 98

net   *noun*

A net is the name given to a flat shape that
will form a three-dimensional object when it
is fixed together in the right way. Nets are
useful for dressmaking, making boxes for
packaging, or for making models.
*He used the net to make a pyramid for the
Egyptian display.*

glue tab
fold
cut

a cube

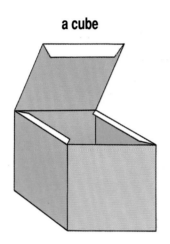

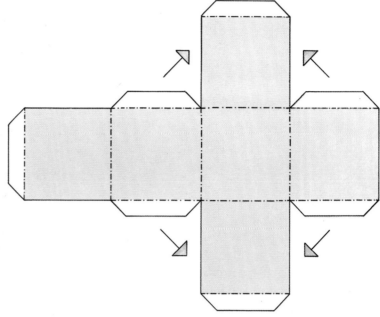

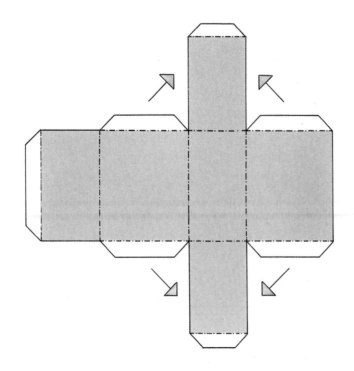

glue tab
fold
cut

a cuboid

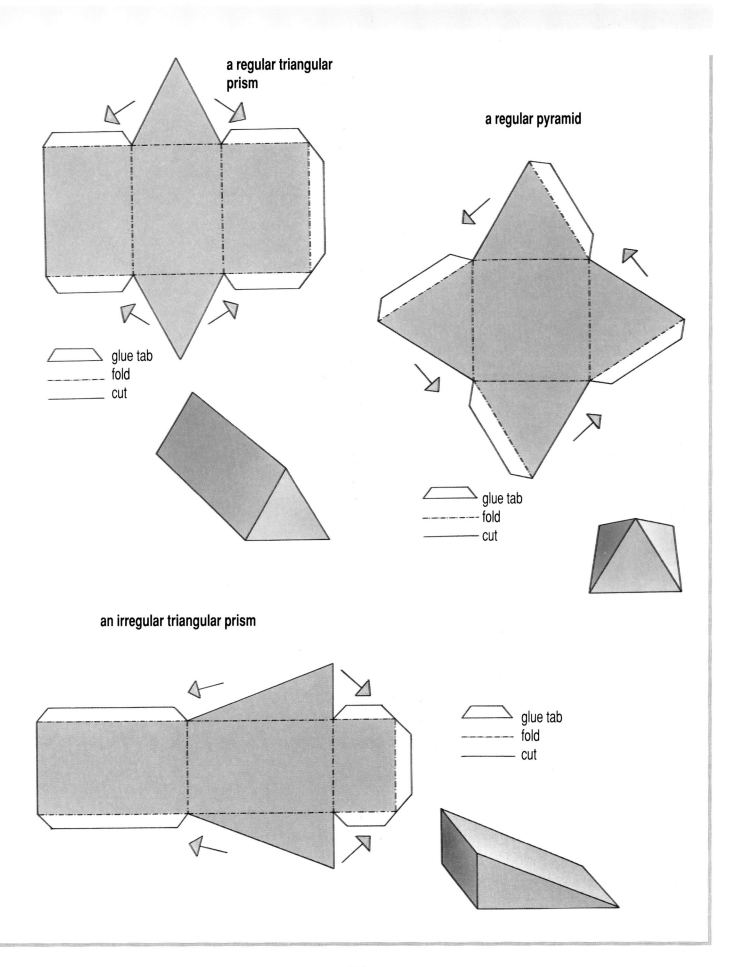

a regular triangular prism

a regular pyramid

glue tab
fold
cut

glue tab
fold
cut

an irregular triangular prism

glue tab
fold
cut

numeral   *noun*

A numeral is a symbol that stands for a number. Numerals are grouped in various ways to make a numeral system, such as the **decimal system**, that enables people to make calculations. People write numerals in different ways in different languages.
The numerals used to number the pages in this book are known as Hindu-Arabic numerals.

Roman numerals are often used on a clock face.

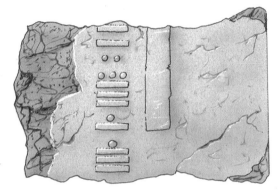

The Mayan Indians wrote groups of numerals vertically, working to base 20.

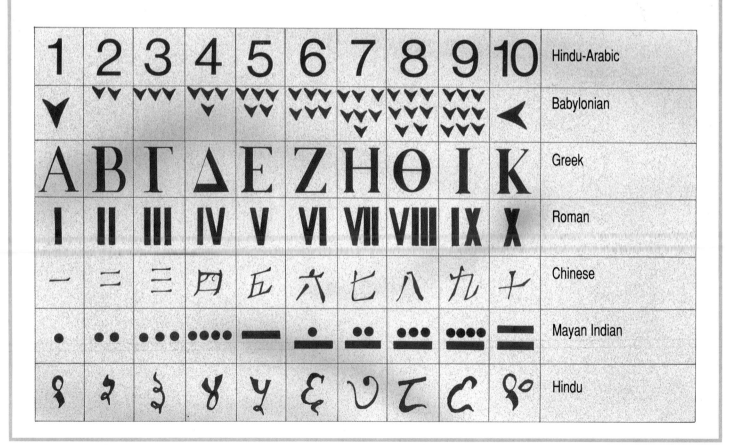

numerator   *noun*
The numerator of a **fraction** is the number above the line or **bar**.
The numerator of the fraction $\frac{4}{7}$ is 4.

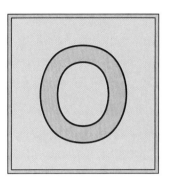

oak tree shilling   *noun*
An oak tree shilling was a coin used in the North American colonies before the United States was created. It was stamped with a picture of an oak tree.

oblong   *noun*
An oblong is another name for a rectangle.
The shape of a football pitch is oblong.

obtuse angle   *noun*
An obtuse angle is an angle that is larger than 90° but smaller than 180°.
The scissors opened fully to make an obtuse angle.

octagon   *noun*
An octagon is an eight-sided polygon.
The interior angles of a regular octagon measure 135°.

octahedron   *noun*
An octahedron is a solid shape that has eight faces.
An octahedron is a polyhedron.

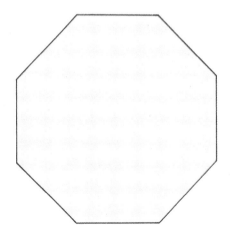

odd number   *noun*
An odd number is a number that cannot be divided exactly by 2. For example, 3, 7, 11 and 23 are odd numbers because when divided by 2 there is a **remainder**. A number that can be divided exactly by two is an **even number**.
The sum of two odd numbers is always an even number.

odds   *plural noun*
Odds are an estimate of how likely it is that something will happen. At a horse race, people can bet on which horse will win. If the odds on one horse are 4 to 1 and the horse wins, those people will win four times as much money as they placed on the bet.
The teacher asked them to work out the odds of rolling eight with a pair of dice.

opposite angles   *plural noun*
Opposite angles are two pairs of angles made when two straight lines cross.
The members of a pair of opposite angles are always the same size.

order   *verb*
To order something means to ask someone to get, or supply it. Bookshops do not keep every book that is published because there is not always always enough space to do so. If a customer wants a book that is not in **stock**, the shop will order it.
The shop ordered 10 more televisions from the wholesale store.

ordered pair   *noun*
An ordered pair is two numbers or symbols which must be written in a certain way to present the correct information. The ordered pair (3, 7) is not the same point as the ordered pair (7, 3). In graph work, **co-ordinates** are ordered pairs. The first number always refers to a position on the **horizontal** axis and the second number refers to a point on the **vertical** axis.
Find the ordered pair 5,H on the map, to see where my street is.

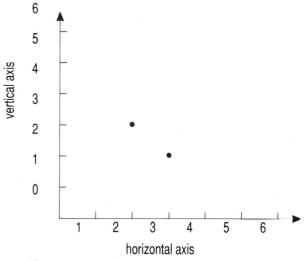

ordinary number ▶ scalar

organization   *noun*
An organization is a group of people who work, meet or play together, usually with some shared interest or common goal in mind.
A company is an organization.

ounce   *noun*
An ounce is an **imperial** measure of weight or volume. There are 16 ounces in one pound. An ounce is equivalent to approximately 28 grams in the metric system. An ounce is also a liquid measure.
The flour weighed 10 ounces.

oval   *noun*
An oval is a kind of curved **shape**. It is the shape of a squashed circle, or **ellipse**.
His drawing of an egg was oval.

overdraft *noun*
An overdraft is a form of **loan**. Customers can ask a bank if they can spend a sum of money above the amount in their account. If the bank agrees, it may charge **interest** on the overdraft.
The bank gave him an overdraft so that he could pay for his holiday.
overdraw *verb*

Pakistani rupee *noun*
The Pakistani rupee is the currency of Pakistan. It is made up of 100 paisas.

paper money ▶ page 102

parabola *noun*
A parabola is a kind of curve. A parabola is the **shape** made if a **cone** is cut **parallel** to its side.
A parabola was drawn to show how the ball travelled through the air.

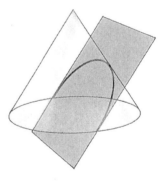

parallel *adjective*
Parallel describes two straight lines which never meet, however long they are. Parallel lines are always the same distance apart. Railway tracks are parallel lines.
They drew a line parallel to the edge of the pavement.

parallelogram *noun*
A parallelogram is any four-sided, or **quadrilateral**, shape in which opposite sides are **parallel**. A **rhombus** is a parallelogram.
Parallelogram shapes were used to form the mosaic.

paper money   *noun*

Paper money is **currency** which is printed on paper. The Chinese were the first people to use paper money instead of coins, perhaps as long ago as 600 AD. **Bank notes**, **bills of exchange** and treasury notes are all different types of paper money. Today, many countries are replacing low **denomination** paper money with coins. The cost of printing paper money and the length of time it lasts makes it more economical to **mint** coins.

Bank tellers keep paper money separate from the coins.

China was the first country to have paper money.

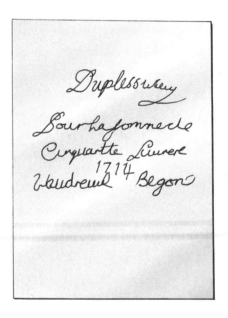

Playing cards were sometimes used as money in Canada in the late 1600s and early 1700s. The card had to be signed by the governor before it could be used as currency.

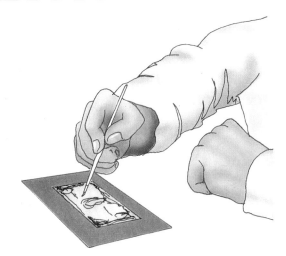

An artist designs the note.

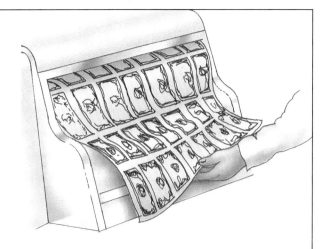

The notes are printed in sheets. Special paper and inks are used.

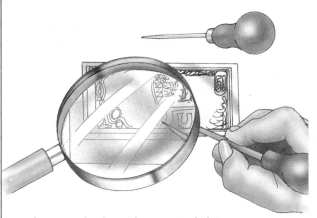

An engraving is made on a steel plate.

The sheets are inspected for flaws, and cut to size.

Every detail must be exact before the plates are used for printing.

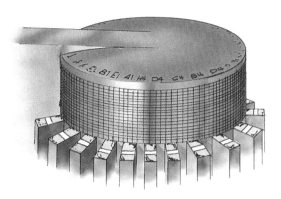

The bills are stacked and counted by machine.

parity   *noun*

Parity means having the same value. Another way of saying two things have parity is to say they are 'at par'. Two currencies are at parity if they are worth the same. For many years, the Irish punt and the British pound shared the same value.
The punt and the pound sterling were at parity.

pattern   *noun*

In geometry, pattern is created when a particular shape or group of shapes is repeated a number of times. The shapes can be repeated by **translation**, or they can be **rotated**. A pattern is often called a design.
A pattern was used to decorate the wall.

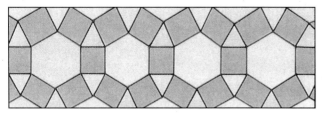

pawnbroker   *noun*

A pawnbroker is a person who lends money in exchange for **goods**. For example, a customer can borrow 50 pounds against a possession which has a higher value. The customer pawns the possession for cash. The pawnbroker will charge **interest** on the money while the goods are held. The pawnbroker will return the goods to their owner only if the money and interest are repaid.
The pawnbroker lent him 100 pounds in exchange for his gold watch.

payment   *noun*

A payment is a sum of money paid for supplying goods or services. A business might make a payment to an electricity company for supplying power. It will also make a weekly or monthly payment to its employees in return for the work they do.
The payments made by the company last month totalled £85,000.

penny *(plural* **pennies***)*   *noun*

A penny is a coin. In the United States, penny means the same as **cent**. There are 100 pennies, or cents, in a dollar. In Britain, there are 100 pennies in a pound sterling.
The child saved up her pennies in a money box.

pension   *noun*

A pension is money paid to people over a certain age. In some countries, people are not allowed to carry on working after they have reached a certain age. As **compensation**, the government pays them a pension. Sometimes, a company also pays a pension to its employees.
The telephonist retired at 50 years of age and collected a small pension.

pentagon   *noun*

A pentagon is a flat **shape**, or polygon, that has five sides.
The interior angles of a regular pentagon are all equal to 108°.
pentagonal   *adjective*

pentagram   *noun*

A pentagram is a star with five points. It can be made by extending the sides of a pentagon with lines until the lines meet. The places where the lines meet are the points of the pentagram.
They used a pentagram as the logo of their company.

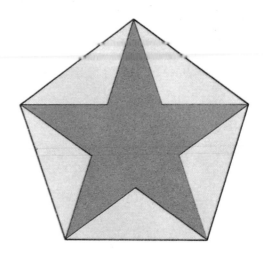

pentomino   *noun*
A pentomino is a shape made by joining five squares together.
A pentomino is a polyomino that can be used in tessellation.

per annum   *adverb*
Per annum means each year. An employee who is paid £15,000 every year is said to earn £15,000 per annum.
The borrower paid interest of £400 per annum.

per cent   *adverb*
Per cent means a fraction, or part, of a hundred. Sixty per cent means sixty hundredth parts or $\frac{60}{100}$. If 60 per cent of people like chocolate, it means that 60 people out of every 100 like chocolate. The symbol for per cent is %. The full term for per cent is percentage.
Seventy-five per cent of the crowd supported the home team.
percentage   *noun*

perimeter   *noun*
The perimeter of a flat shape is the distance around it. For example, the perimeter of a hexagon is worked out by adding together the lengths of all six sides.
They walked around the perimeter of the playing field.

permutation   *noun*
A permutation is a rearrangement of a **set** of things in a different order. The set of numbers 1 3 5 7 9 is one permutation of 3 7 5 1 9.
They changed the permutation of their combination lock on the safe.

perpendicular   *adjective*
Perpendicular describes lines that meet at 90°. Lines that are formed at **right angles** to a base line, are said to be perpendicular. **Vertical** lines are perpendicular to **horizontal** lines.
The mast rose perpendicular to the deck.

peseta   *noun*
The peseta is the currency of Spain.

peso   *noun*
The peso is the unit of currency of several countries of the world, including Colombia, Cuba and the Philippines. It is divided into 100 cents or centavos.

petty cash   *noun*
Petty cash is a small amount of money held by a business. It is used to buy small items such as postage stamps.
Last week, the business spent £23 from petty cash.

pi   *noun*
Pi is a number that equals the **circumference** of a circle divided by its **diameter**. It is a **ratio** which is the same for any circle. It is calculated as 3.14159 to six **significant figures**.
The symbol for pi is π.

pictogram   *noun*
A pictogram is a **graph** which uses small symbols or pictures to represent information.
They recorded their household pets on a pictogram.

pie chart *noun*
A pie chart is a graph which presents information using a circle that has been divided into **sectors**. Each sector represents a **fraction** of the whole pie, or circle.
The pie chart showed how many visitors came from each European country.

piece of eight *noun*
A piece of eight was a coin once used in Spain and its colonies. A piece of eight was worth eight rials.

piggy bank *noun*
A piggy bank is a small, hollow model of a pig. Money is put in through a slot in the top to encourage children to save.
She had four pounds in her piggy bank.

pine tree shilling *noun*
A pine tree shilling was a coin used in the North American colonies, before the United States was created. It had the image of a pine tree stamped on it.

pint *noun*
A pint is a measure of **capacity** in the **imperial measure system**. It is equal to half a quart or one eighth of a gallon. One British pint is equal to 0.568 of a metric litre.
They each drank a pint of milk.

place value *noun*
Place value is a phrase that describes how much a **digit** is worth according to its position in a number. In the number 372, the 3 is worth 300, the 7 is worth 70 and the 2 is worth 2. If these are added together, the whole number is worth 300 + 70 + 2 = 372.
The place value of 6 in 62 is worth 60.

plane *noun*
A plane is a flat surface, such as a sheet of paper. It is **two-dimensional**. Any point on the plane needs two **co-ordinates** to locate it, those of length and width.
The machine had to be positioned on a horizontal plane so it would not move.

plane figure *noun*
A plane figure is a **shape** with only two dimensions.
A square is a plane figure but a cube is a solid.

plot *verb*
To plot means to mark points on a chart, grid or graph. When results are plotted, they can be written as **co-ordinates** on a graph or chart, so that information can be read off more easily.
The navigator plotted a course on his map.

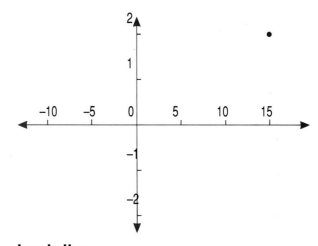

plumb-line *noun*
A plumb-line is a length of string attached to a weight. The weight pulls the string downwards and holds it straight. The plumb-line can then be used to check whether an edge or line is **vertical** to the base line.
They used a plumb-line to check that the edge of the wall was vertical.

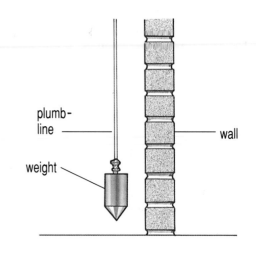

plus *preposition*

Plus is the word used for **addition**. Plus can also be represented by the **symbol** +.
We worked out that six plus eight came to fourteen.

PM *abbreviation*

PM or pm stands for the Latin words, 'post meridiem', which mean 'after mid-day'. PM is used to indicate time that falls in the afternoon and evening between 12 o'clock mid-day and 12 o'clock midnight.
Her train was scheduled to depart in the evening at 6 pm.

pocket money *noun*

Pocket money is an allowance that parents give their children. The children can usually spend the money as they like, or they can save it.
She got £5 pocket money each week which she spent on her pets.

point *noun*

1. A point is a position on a map or **graph**.
He knew where the point was by its co-ordinates.
2. A point is a dot that comes after the whole number in a decimal number. The decimal number 3.4 means $3\frac{4}{10}$. The decimal point tells us that the whole number is 3 and that the 4 stands for 4 parts of 10.
In 34.67, the decimal point comes after the 34 whole number.
3. A point is a place where two lines meet, such as the sharp corner at the tip of one arm of a star.
She counted five points on the picture of the star.

point of sale *noun*

A point of sale is a place in a shop where goods are paid for. Many large shops install computer equipment at the point of sale. Each item has a **bar code** which is used to register the price of the item in the computer.
She paid for the gardening book at the point of sale near the exit to the store.

policy *noun*

1. A policy is a decision, often made by a business, that it will carry out its activities in a particular way.
It is the company's policy to pay bonuses to staff at Christmas.
2. A policy is an agreement between an **insurance** company and a **client** that the company will pay out money as **compensation** if certain things occur. In most cases, these occurrences refer to theft, injury or death. The policy is written down in a document which states what must occur before the company will pay out any money to the client.
The policy said that £100,000 would be paid if the customer is injured.

polygon *noun*

A polygon is a flat **shape** with edges that are straight lines. If all the edges are the same length and all the interior angles are the same size, the polygon is called a regular polygon.
A square table top is a regular polygon.

polyhedron (plural **polyhedra**) *noun*

A polyhedron is a **solid shape** with many flat surfaces called faces. A dodecahedron, an icosahedron, and a tetrahedron are all examples of polyhedra.
They drew a polyhedron with 10 faces.

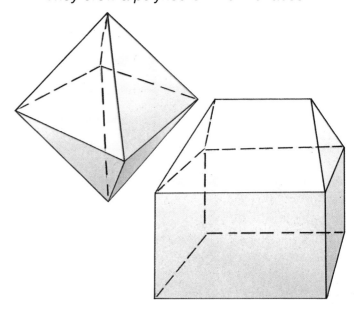

polyomino *noun*

A polyomino is a flat shape made by joining squares together. Polyominos have special names depending on the number of squares that are joined. Polyominos can be used in **tessellation**.

The face of the table was a polyomino made of wooden squares.

Name of polyomino	Number of squares
Domino	2
Tromino	3
Tetromino	4
Pentomino	5

population *noun*

Population is a word used in **statistics** to describe a group or **set**, of things from which a sample has been taken. In a factory, one sample may be inspected from every 100 products manufactured, or the product population.

They tested one chocolate from a population of 1,000.

portfolio *noun*

A portfolio is a list of **investments** owned by a company or an individual. For example, a property company may have a large portfolio of houses which it owns.

The value of the shares in the portfolio was 2 million pesos.

positive number *noun*

A positive number is a number that is greater than zero. Adding a positive number to another number always gives a number of higher value. The symbol + is put in front of a positive number when both positive and **negative numbers** are used.

The two positive numbers were added together to obtain a second, greater positive number.

post meridiem ► PM

post office *noun*

A post office is an organization which receives and delivers letters or parcels. It receives payment through sales of postage stamps of various values which must be fixed to the letter or package. The post office may also act as a savings bank.

They purchased three first class stamps from the post office.

postal order *noun*

A postal order is similar to a **cheque**, but it is issued by a **post office** not a **bank**. The recipient can cash the postal order at any post office. Postal orders are also a useful way of mailing money to someone far away.

He paid for the goods by sending a postal order for £12.

pound *noun*

The pound is the unit of currency of several countries of the world, such as Lebanon, Sudan and Syria.

pound *noun*

A pound is an **imperial** measure of weight. It is written lb in its short form. It is equal to 453.59 grams in metric measurement.

A pound of sugar cost 70 pence.

pound sterling *noun*

The pound sterling is the currency of the United Kingdom. It is made up of 100 pence.

poverty *noun*

Poverty describes the situation of people when they do not have enough money to buy the things they need.

The family lived in poverty without sufficient food or clothing.

power ▶ page 110

premium   *noun*
1. A premium is an amount of money that is paid for **insurance**. The premium is usually paid every month or every year.
His premium was £45 per month.
2. A premium is a price greater than the suggested price. A buyer may be forced to pay a premium to purchase goods which are very scarce or highly valued.
The car was priced from £7,000 but fetched a £1,000 premium because it was so rare.

price   *noun*
The price of an item is the amount of money that must be paid to purchase it. A price can be altered by offering a **discount** or by adding a **premium**. It is sometimes possible to **bargain** a price.
The price of the car was $25,000.

price war   *noun*
A price war occurs when two or more companies compete on price to sell goods. The price war begins when one company lowers its prices below those of its competitor to attract more **customers**. The second company then lowers its prices even further. If this pattern continues, customers can benefit from lower prices, but price wars often cause businesses to cut back on **production** or to fail.
The petrol companies entered a price war which drove the cost of petrol down.

prime number   *noun*
A prime number is a number that can only be divided by 1 or by the number itself. For example, 7 is a prime number because its only **factors** are 1 and 7.
Every prime number bigger than 2 is an odd number.

prism   *noun*
A prism is a **solid** shape. It has two ends which are **polygons** of the same size and shape. Its sides are **parallelograms** or **rectangles**.
A glass prism will turn white light into a rainbow.

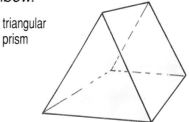

triangular prism

private company   *noun*
A private company is any company that is not a **public company**. A private company may be small, and owned by just a few **shareholders**. Its shares cannot be bought and sold by the public.
The business is a private company owned by a husband and wife.

probability   *noun*
Probability is the likelihood of something happening. It is the **ratio** between one result and the total number of possibilities. The probability of throwing a 4 with a **die** is 1 in 6, or 1:6, because a die has six sides, only one of which shows the number four.
The probability of a coin landing to show heads not tails is calculated as 1 in 2.

problem   *noun*
A problem is a mathematical puzzle. The solution to a problem is found by using one of the many mathematical rules or **formulae**.
The problem was solved using Pythagoras' theorem.

power   *noun*

A power is the result, or product, of a number multiplied by itself. For example 5^2 is 5 to the power of 2, or $5 \times 5 = 25$, so 25 is the second power of 5. $5^3 = 5 \times 5 \times 5$ and the product, 125, is the third power of 5.

Exponents are used to show how often the number is to be multiplied.

They worked out that 100 was the second power of 10.

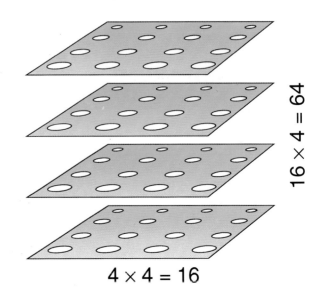

$$16 \times 4 = 64$$

$$4 \times 4 = 16$$

4 multiplied by itself 3 times is 4 to the power of 3 and written 4^3. $4 \times 4 \times 4 = 4 \times 4 = 16 \times 4 = 64$.

4 multiplied by itself zero times remains 4.

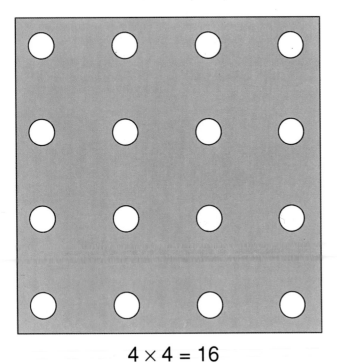

$$4 \times 4 = 16$$

4^2 is four to the power of 2, or 4×4. The number of 4s in the statement tells you the power.

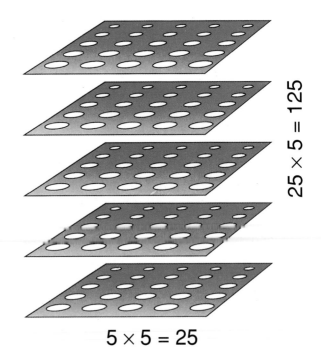

$$25 \times 5 = 125$$

$$5 \times 5 = 25$$

$5 \times 5 \times 5$ is 5 to the third power or 5^3. If you add up all the dots there are 125.

product *noun*

1. A product is something that a business makes or sells. The products of a bakery are all the cakes, biscuits and loaves of bread which it offers for sale.
Companies often publish a list of their products.
2. A product is the number produced when any group of numbers is multiplied together. The product of 3 × 4 is 12.
They worked out the product of 3 × 5 to be 15.

production *noun*

Production means the process of making something. Production can also refer to the quantity of things that are made.
The production of cars increased in July.

profit *noun*

A profit is the money earned when the **income** of a business is greater than its **expenditure**. A business tries to earn as much income as it can, but the business will also have to pay out money to buy **goods** and **services**. If the income is greater than the payments, it makes a profit, sometimes called a gain. The opposite of profit is **loss.**
The company made a profit of £25,000 last year.

profit and loss *noun*

Profit and loss is a type of record of **accounts** of a business. It shows how much **net** profit a business has made in a given period or how much net loss it has suffered.
The profit and loss account showed that the company had made a net profit of £1 last year.

projection *noun*

1. A projection is a map of the Earth's surface. The Earth's surface is curved, so when it is drawn on a piece of paper, some areas are twisted, or distorted. Different projections try to make up for this.
With a Mercator projection, the Earth is drawn as a rectangle.
2. A projection is a kind of **forecast** or estimate, about the future. A company will often try to forecast the number of products it expects to sell in the coming year. To do this it will use information about how many products have been sold in the last few months or years.
The sales projection was that 150 videos would be sold.

property (plural **properties**) *noun*

Property is something that is owned by an individual, a group of individuals or a company. It is often used to describe land or buildings. For example, people's houses are sometimes referred to as their property.
He bought a property near the park.

proportional ▶ direct proportion

protractor *noun*

A protractor is an instrument for measuring angles. It is usually shaped in the form of a **semicircle**, but a circular protractor is useful for measuring **bearings** and certain other kinds of angle. Protractors are marked on the curved edge to show the number of degrees the angle is turning.
She used the protractor to measure the angles of the triangle.

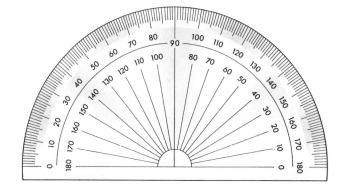

public company   *noun*
A public company is a company with **shares** that can be bought and sold by members of the public, usually on a **stock exchange**. Public companies are usually large, and have many **shareholders**.
The public company decided to sell more shares to the public to raise cash for a new project.

purchase   *verb*
To purchase means the same as to buy.
He purchased a garden shed.

pure mathematics   *noun*
Pure mathematics describes the study of mathematical problems for their own sake. They are not studied in order to provide solutions which will be put to some practical use. The opposite of pure mathematics is **applied mathematics**.
The college offered a course in pure mathematics.

puzzle   *noun*
A puzzle is a kind of problem. In a jigsaw puzzle, the problem that is set is to fit all the pieces together properly.
A tangram is a puzzle.

pyramid   *noun*
A pyramid is a solid shape. It has a flat **polygon** base and its sides rise up to a point or **apex**, at the top. Each of the sides is a triangle.
The Ancient Egyptians built tombs for their kings in the shape of a pyramid.

Pythagoras' theorem   *noun*
Pythagoras' theorem describes how the lengths of the sides of a right-angled triangle are related to each other. It states that the **square**, meaning squared length, of the **hypotenuse** is equal to the square of the other two sides. Pythagoras' theorem is written as $c^2 = a^2 \times b^2$, where c is the length of the hypotenuse. Pythagoras' theorem can be worked out by drawing squares on the edges of the triangle. The **area** of the square on the hypotenuse is equal to the sum of the areas on the other two sides.
Pythagoras' theorem was used by the Ancient Egyptians when they built the pyramids.

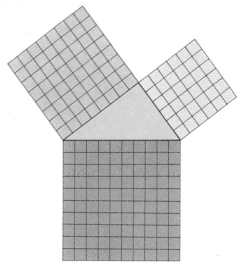

quadrant   *noun*
A quadrant is one quarter of a **circle**. It has the shape of a **sector**. The angle at the centre of a quadrant is always a **right angle**, measuring 90°.
We cut the pie into four quadrants to make four equal shares.

quadrilateral   *noun*
A quadrilateral is a **polygon** that has four sides and four angles. There are many kinds of quadrilateral. Rectangles, parallelograms, rhombuses, trapezoids and diamonds are all quadrilaterals.
The internal angles in a quadrilateral add up to 360°.

quadruple   *verb*
To quadruple means to **multiply** a number by four.
The sales figures should quadruple this year.

quantity   *noun*
A quantity is an amount. Quantities can be described in numbers.
After selling 250 books in one morning, the quantity of books in the shop was severely decreased.

quart   *noun*
A quart is an **imperial measure** of **capacity** equal to two pints or a quarter of a gallon. One British quart equals 1.136 litres in the metric system. One American quart is equal to 0.946 litres.
They drank a quart of orange juice at the party.

quarter   *noun*
1. A quarter is one part of something that has been divided into four equal parts. It is one fourth of something. One quarter can also be written $\frac{1}{4}$.
A quarter of 12 is 3.
2. A quarter is an American or Canadian coin worth 25 cents. It is a quarter of a dollar.
3. A quarter is also an imperial measure of weight. There are 28 **pounds**, or lbs, in a quarter.
The grocer bought a quarter of pears.

quetzal   *noun*
The quetzal is the currency of Guatemala. It is made up of 100 centavos.

quipu   *noun*
A quipu was a set of coloured strings with knots. It was used in ancient Peru for counting and also as a form of calendar.
He counted the amount of gold on the quipu.

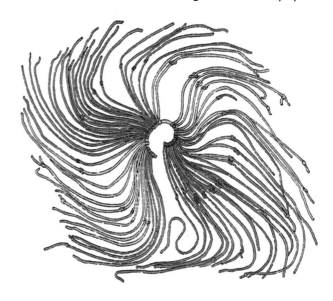

quotient   *noun*
The quotient is the result which is obtained when one number is divided by another.
He found the quotient of the division sum 28 ÷7 was 4.

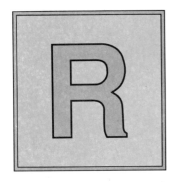

radius (plural radii) *noun*

The radius of a circle is the distance from the centre of the circle to the edge, or circumference. This distance is always half the length of the **diameter**.
The radius of the circle that she drew was 5 centimetres.

rand *noun*

The rand is the currency of South Africa. A rand is 100 cents. South Africa also has a specially minted gold coin called a krugerand.

random *adjective*

Random is used to describe things that happen by chance. When things are chosen at random, it means that they are chosen without making a plan or without any particular intention.
The team members were chosen at random by picking names from a hat.

rate *noun*

1. A rate is a number or value that describes how many units of one thing occur in something else. The speed of a car is a rate because it measures how many kilometres are travelled in each hour.
The crowd entered the stadium at a rate of 2,000 each hour.
2. A rate is an amount of money charged for a service. For example, an accountant might charge for services at a rate of $50 per hour. A bank might lend money and charge a rate of interest of 10 per cent per annum.
The rate of pay for this employee is 40 francs per hour.

rate of exchange *noun*

The rate of exchange is the price or value at which one **currency** is exchanged for another. It is usually different depending on whether a currency is being bought or sold, and can vary according to where the **transaction** takes place. The rate of exchange varies daily as the value of one currency moves against another according to demand.
The rate of exchange for the pound in New York was $1.50 on Saturday, but it rose to $1.53 on Monday.

ratio ► page 116

ration *noun*

A ration is a limited amount of something. For example, if petrol becomes scarce, it may need to be shared out equally to everyone who has a car. Each share is called a ration.
The customers had to queue to receive a ration of meat.

rational number *noun*

A rational number is a number that can be written as a **fraction**. For example, 0.625 is a rational number because it can also be written as $\frac{5}{8}$. Rational numbers include all whole numbers because they can easily be written as a fraction with a denominator of 1.
9 is a rational number when it is written as the fraction $\frac{9}{1}$.

real *noun*
A real is a silver coin that was once used in Spain and its colonies.

real estate *noun*
Real estate is a term that means buildings and the land they are built on.
The real estate on the holiday coast was worth $25,000,000.

receipt *noun*
A receipt is a piece of paper that is given to customers when they buy something. It proves where the item was bought and how much it cost.
The receipt proved that the customer bought the radio from the electrical store.

receiver *noun*
A receiver is a person who takes over responsibility for the property of a **bankrupt** company. The receiver sells as many **assets** as possible, in order to raise as much money as possible to pay the company's debts.
The bank appointed a receiver to a company with one million pounds of debts.

receivership *noun*
Receivership is the condition a company is in when its **assets** are under the control of a **receiver**.
The bicycle company went into receivership when its debts became too great to pay.

recession *noun*
A recession is a time when the **economy** is not very active. Companies find they cannot sell as many goods as before. As a result, manufacturing companies also slow down production. Companies need fewer staff, and may have to make employees **redundant**.
Europe and the United States have suffered from a period of recession in recent years.

reciprocal *noun*
The reciprocal of a number is 1 divided by the number. The reciprocal of 4 is $\frac{1}{4}$. If a number is multiplied by its reciprocal, the answer is always 1. For example, $4 \times \frac{1}{4} = 1$.
The reciprocal of 5 is $\frac{1}{5}$.

rectangle *noun*
A rectangle is a **polygon** that has four sides. Each of the sides meets at a right angle. The opposite sides of a rectangle are always equal.
Each face of a brick is a rectangle shape.
rectangular *adjective*

rectangular number *noun*
A rectangular number is a number that can be arranged as a rectangular pattern of dots. Six is a rectangular number. Seven is not a rectangular number. All **even numbers** above 2 are rectangular numbers.
No prime number can be a rectangular number.

reduce *verb*
To reduce means to make smaller or less. Reducing is often used when working out **percentages**.
The shop reduced the price by 10%.

redundant *adjective*
Redundant describes things or people that are no longer needed. If a clothes factory cannot sell its clothes, it may have to make some of its workers redundant because it cannot afford to employ them.
Sixty factory workers were made redundant.

ratio *noun*

Ratio is a method of comparing two quantities. Ratios can also describe the rate at which something is happening, such as kilometres per hour, or the **probability**, or likelihood, of something happening. At a football match, if there are 300 supporters of the home team and 200 supporters of the visiting team, the ratio is 300 to 200 or 3:2. In other words, for every three home supporters there are two visiting supporters. *In the traffic survey the ratio of cars to buses was 15 to 1.*

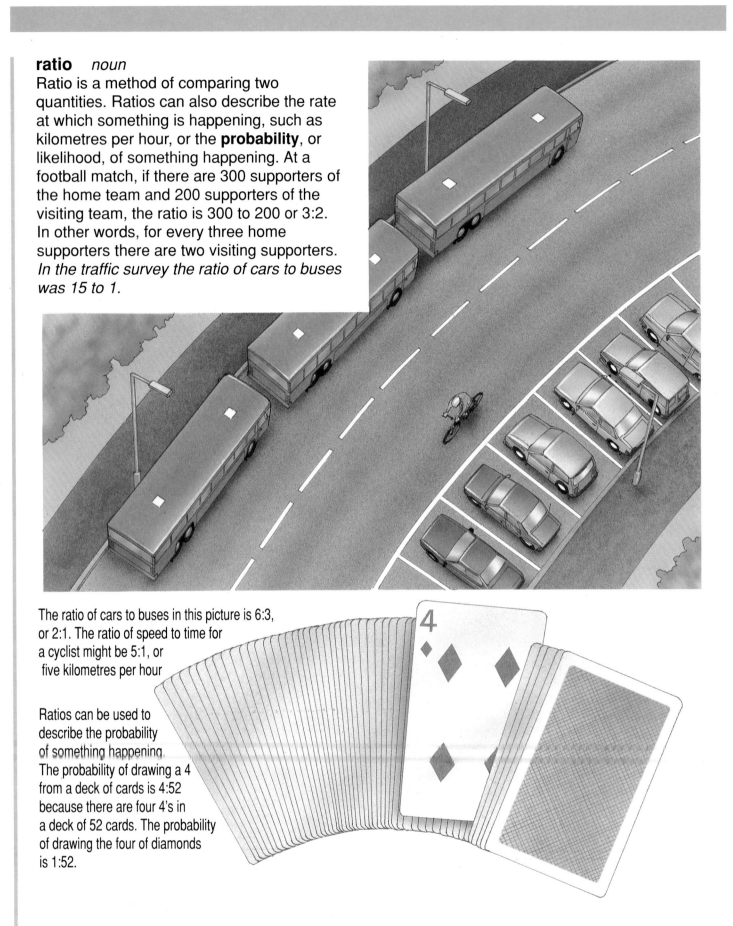

The ratio of cars to buses in this picture is 6:3, or 2:1. The ratio of speed to time for a cyclist might be 5:1, or five kilometres per hour

Ratios can be used to describe the probability of something happening. The probability of drawing a 4 from a deck of cards is 4:52 because there are four 4's in a deck of 52 cards. The probability of drawing the four of diamonds is 1:52.

the golden ratio

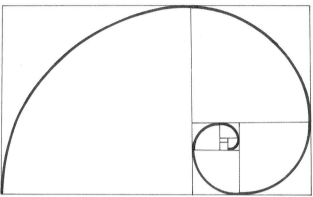

When a square is taken away from a golden rectangle, another golden rectangle is left. This can be repeated many times.

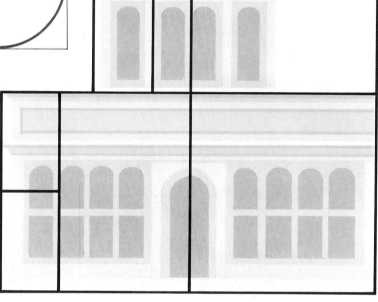

The golden ratio is also known as the golden section or divine proportion. It is approximately 1:1.618. A rectangle, the length and width of which match this ratio is very pleasing to the eye, and is called a golden rectangle.

The golden ratio has been applied to many buildings.

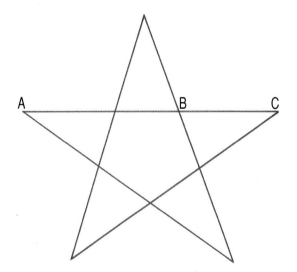

The ratio of AB to BC on a pentagram is the golden ratio.

The golden ratio can often be found in nature.

reflection *noun*
A reflection is a form of **symmetry**. If a line is drawn down the middle of an **isosceles triangle**, the side on the left of the line is an exact copy, or reflection, of the side on the right. A mirror placed down the centre would show that each side and its reflection look like the whole triangle.
Each side of the playing card is a reflection of the other side.

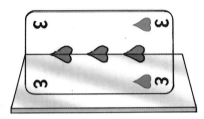

reflex angle *noun*
A reflex angle is an angle that is bigger, or greater than 180° but smaller, or less than 360°. At ten minutes to two, the angle between the hour hand and the minute hand is a reflex angle.
When she opened the fan too far, the two arms made a reflex angle.

regular shape *noun*
A regular shape is one in which all the sides are of equal length and all the angles are of equal size.
A square is a regular shape.

remainder *noun*
A remainder is any number that is left over when one number is divided by another. r is the symbol for a remainder.
Fourteen divided by four gives an answer of three with a remainder of two.

remuneration *noun*
Remuneration is a payment made to someone for doing a job.
His remuneration for painting the house was £2,000.

Renminbi Yuan *noun*
The Renminbi Yuan is the currency of China. It is divided into 10 jiao.

rent *noun*
Rent is a payment made for the use of an **asset**, often an expensive piece of equipment, a car or a building. It is paid to the person who owns the asset. Rent is usually paid at regular intervals, perhaps each month or each year.
Some companies pay rent for the use of a factory.
rent *verb*

repay *verb*
Repay means to pay back money that has been borrowed.
He had to repay the loan to the bank in three years.

reserve *noun*
A reserve is an amount of money that is set aside until it is needed for some particular purpose.
He had a reserve of a thousand pounds in case of emergency.

resolution *noun*
A resolution is the result of solving a problem. It can also mean the process of solving a problem.
They could not find the resolution to that problem.

result *noun*
A result is an answer, or outcome, obtained when a problem is solved.
The result of the problem proved the theory to be true.

retail ► page 120

retail price index *noun*

The retail price index is a number, or **index**, used to indicate whether the retail prices of goods being sold in stores throughout a country, are going up or down.

The retail price index showed that prices of electrical equipment had dropped during the year.

return ▶ yield

revenue ▶ income

revolution *noun*

A revolution in mathematics refers to a full turn of 360°. The second hand on a clock face turns through one revolution every 60 seconds. Objects and diagrams are sometimes revolved in **geometry** to test them on different **co-ordinates**, **grids** or **axes**.

A car wheel makes one revolution every time it spins.

rhombus *noun*

A rhombus is a **parallelogram** that has all its sides of equal length. Two of its **opposite angles** are **obtuse angles** and the other two are **acute angles**.

A rhombus is a diamond shape.

rhomboid *adjective*

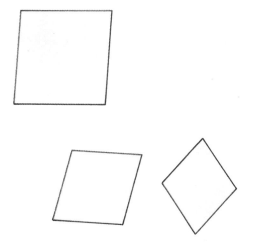

rial *noun*

The rial, sometimes spelt riyal, is the unit of currency of several Middle East countries, including Iran, Oman, Qatar and Saudi Arabia.

right angle *noun*

A right angle is an angle that measures exactly 90°.

The angles in a square are right angles.

right-angled triangle *noun*

A right-angled triangle is a triangle in which one of the angles is 90°. The sum of the other two angles is always 90°. These triangles are sometimes called right angles.

The 90° angle in each of these right-angled triangles is identified by a red box.

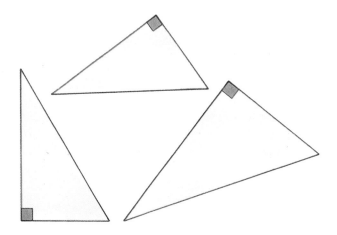

ringgit *noun*

The ringgit is the currency of Malaysia. It is divided into 100 sen.

risk *noun*

A risk is the possibility of taking a wrong decision and losing money in business. A company might buy an expensive machine to make a new kind of product. If the company cannot sell enough of the product, the money it spent on the machine will be wasted.

The directors thought carefully about the risk before they bought the machine.

riyal ▶ rial

retail *verb*

To retail is to sell to an individual, usually in a shop. Retail is different to **wholesale**. Wholesale goods are only sold to other businesses in bulk. Retailing involves everything necessary to operate a shop, from serving customers to ordering and displaying goods.

The flower shop also retails flowers from a stall in the market.

retail adjective

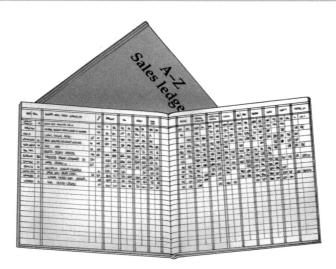

A sales ledger records each transaction. Today, this information is often held on computer.

Money comes into the store in payment for goods sold, and is deposited in the bank.

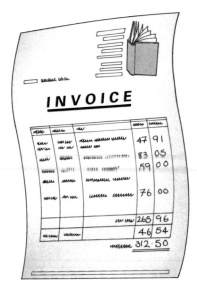

Invoices from suppliers must be paid.

This shop is a retail book shop.

Advertising on television helps to boost sales.

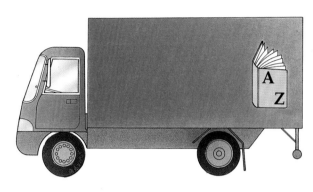

Books must be ordered and delivered to keep the shelves stocked.

The goods sold by the shop are displayed in the window.

Packaging carrying the book shop's logo is a kind of advertising.

The bar code on the back of the book gives coded information that can be recorded on computer.

Roman numerals   *plural noun*

Roman numerals are letters that were used as numerals by the Romans in ancient times. They were later replaced by **decimal** numbers.

Some clock faces still use Roman numerals to indicate the hours.

Roman number	Decimal number
I	1
II	2
III	3
IV	4
V	5
X	10
L	50
C	100
D	500
M	1000

Roman numerals
for 1999

root   *noun*

A root is a number that is multiplied by itself. A square root is a number that is multiplied by itself once, and a cube root is multiplied by itself twice. Another name for root is **power**.

The fourth root of 16 is 2 because 2 × 2 × 2 × 2 = 16.

rotate   *verb*

Rotate is a word that means to turn. Shapes are rotated in geometry when using **symmetry**.

The Earth rotates on its axis once every 24 hours.

rouble   *noun*

The rouble was the currency of the country formerly called the Soviet Union. It is still the currency of the Russian Republic. There are 100 kopecks in a rouble.

round   *verb*

To round is to change a number by increasing it or decreasing it to the nearest ten, hundred or thousand. If the answer to a problem is 2,325, the answer rounded to the nearest hundred is 2,300. A number like 2,372 would be rounded upwards to 2,400 because 2,372 is nearer 2,400 than it is to 2,300. A number that is half way between two hundreds is always rounded upwards, so 2,350 would be rounded to 2,400.

Some decimal fractions need rounding because they go on for ever.

round number ► round

royalty   *noun*

A royalty is a payment made to the owner of an invention for its use. A royalty might be paid to the author of a novel by the publisher, for each copy of a book sold. A royalty might also be offered to a mine owner for the right to work the mine. A royalty normally represents a **percentage** of the **profit** or the sale price of an item.

The author received a royalty of seven per cent on the sales from his first novel.

ruble ► rouble

ruler   *noun*

A ruler is a strip of metal, plastic or wood with a straight edge. It is used when drawing a straight line with a pencil or other writing instrument. A ruler is usually marked in **centimetres** or **inches**.

The student used a ruler to measure the piece of card.

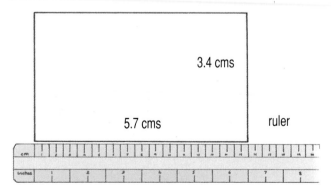

3.4 cms

5.7 cms

ruler

rupee *noun*
The rupee is the unit of currency of several countries of the world, such as India, Nepal, Pakistan and Sri Lanka. In India and Pakistan, the rupee is divided into 100 paise, and in Nepal, into 100 pice. In Sri Lanka the rupee is worth 100 cents.

rupiah *noun*
The rupiah is the currency of Indonesia. There are 100 sen in one rupiah.

safe ► page 124

safe deposit box *noun*
A safe deposit box is usually found in a bank. The bank holds it on behalf of a customer for their private use. Valuables, such as money, **documents**, ornaments or jewellery, are examples of things that might be stored in a safe deposit box.
The diamond ring was put in a safe deposit box at the local bank.

salary *noun*
A salary is a fixed amount of money that an **employee** is paid to do a job. Another word for salary is **wages**.
His salary for last month was $2,500.

sale *noun*
1. A sale takes place when goods are supplied to someone in exchange for money.
He had to reduce the price to make a sale.
2. A sale occurs when a company offers its goods at lower prices than usual. Shops often do this to clear out old **stock** and make space for new goods.
The department store had a sale during January.

sales tax *noun*
Sales tax is a special tax paid by customers when they buy goods. It is usually a **percentage** of the price of the goods. In Britain, sales tax is called VAT, or value added tax. In other countries it is given a different name.
Sales tax in the state was raised to five per cent.

safe *noun*

A safe is a fire-proof and burglar-proof room, or a strong box that is fastened by a door with a lock on it. Banks use safes to hold money and other valuable items. Safe deposit boxes are located in a special sealed room, or vault. Nobody can go in unless they are accompanied by a bank guard or official.

The bank locked the safe at 5 o'clock.

This type of safe is often found in a branch bank. The door is very thick.

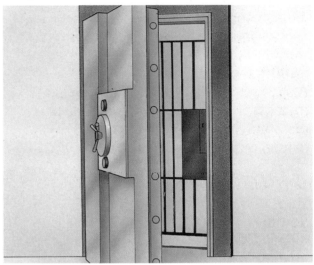

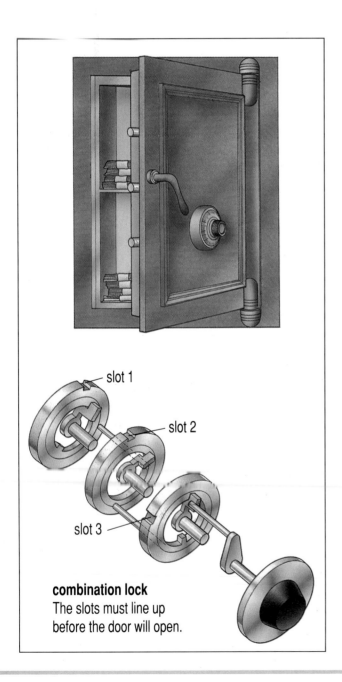

slot 1

slot 2

slot 3

combination lock
The slots must line up before the door will open.

Safe deposit boxes are rented by people who want to protect their valuables. The boxes may be small drawers or larger safes.

savings   *noun*
Savings are amounts of money that are kept and not spent. People often pay the money into a **bank** or **building society** where it can earn **interest**. They can withdraw the money if they need it.
Her savings amounted to £500.

savings bond   *noun*
A savings bond is a kind of **investment**. **Interest** is paid on the money used to buy the bond. After a period of time, the bond can be exchanged for the price paid plus interest.
She paid £200 to buy a savings bond.

scalar   *noun*
A scalar is a kind of number called an ordinary number. The word scalar is used when it is important not to get ordinary numbers mixed up with **vectors**. Ordinary numbers show size but not direction. They are used to indicate measurements like temperature, length or time.
Scalars were used to compare the lengths of the different planks.

scale   *noun*
1. A scale is a set of marks on a line used for measuring. The scale of a ruler shows millimetres and centimetres. The scale of a **protractor** shows **degrees** of angle, while the scale of a **thermometer** shows degrees of temperature.
She saw that the temperature was 20° Celsius on the thermometer scale.
2. A set of scales is a machine used for weighing.
The chef weighed the flour on the scales.
3. A scale is used on a map or drawing to make sure the **area** being drawn is in **proportion** to the real thing. A small measurement on paper represents a larger one in real life. For instance, one centimetre on a map might represent one kilometre across land.
The scale of the map was one centimetre to one kilometre.

scalene triangle   *noun*
A scalene triangle is a triangle in which all the sides are different lengths.
In a scalene triangle all the angles are different.

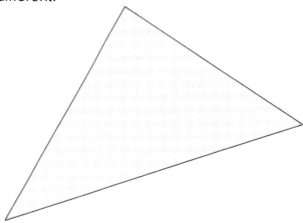

schilling   *noun*
The schilling is the currency of Austria. The schilling is made up of 100 groschen.

score   *noun*
1. A score is an amount owed. It is a debt.
They saved all their pocket money so they could settle the score.
2. A score is a record of points made in a game.
The score in the game of football was three goals to one.

second   *noun*
1. A second is a unit of time. There are 60 seconds in a minute.
The athlete succeeded in running 100 metres in 14 seconds.
2. A second is a unit of angle. There are 60 seconds in a minute of angle.
The angle of the corner of the room measured 90 degrees and 30 seconds.

sector   *noun*
A sector is part of a circle. It is the space between two **radii** and the **circumference** of the circle. A sliced wedge of pie is shaped like a sector.
They used sectors of the circle to make a pie chart.

securities *noun*
Securities is the name given to describe many kinds of **stocks** and **shares** which are offered for sale.
Their securities included ordinary shares and some stocks in four different companies.

segment *noun*
1. A segment is part of a line.
He measured the line segment from A to D.
2. A segment is part of a circle. It is the space between a **chord** and the **circumference** of the circle. A **semicircle** is a large segment of a circle.
The circle was divided into two unequal segments.

sell *verb*
Sell means to supply goods or services in exchange for money.
A bookshop sells books.

semicircle *noun*
A semicircle is a shape made by dividing a circle in half.
When the fan was fully opened, it made a semicircle.

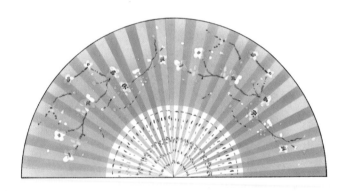

sequence *noun*
A sequence is a **set** of things arranged in a particular order. The **digits** in the number 53864 make a sequence. If this set of digits is arranged in increasing order they form a different sequence, 34568.
A binary number is a sequence of 0's and 1's.

serial number *noun*
A serial number is a number that is used to describe one particular item. Serial numbers help companies to keep track of the things they make. A company that makes CD players will give a different serial number to each machine, so that the company knows exactly which and how many have been sold.
The police asked the owner to identify his stolen radio by its serial number.

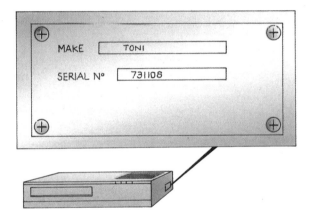

series *noun*
A series is a **sequence** of numbers that are in some way linked to follow a pattern. For example, in the series 7 11 15 19 it can be seen that the numbers form a pattern because four is added to each new number. So, the next numbers in the series will be 23 and 27.
They worked out the first two numbers in the pattern by recognizing how the sequence was formed.

service *noun*
A service is a job that is carried out for customers or clients. For example, a motor mechanic might repair a car, or an accountant might prepare accounts and advise about tax. Service businesses sell services rather than **goods**.
The company paid for the services of a lawyer.

set *noun*
A set is any collection of things that have something in common. For instance, a group of apples in a basket would be a set, as would any group of numbers or objects with something in common.
The number six belongs to the set of even numbers.

set square *noun*
A set square is a triangle made out of wood or plastic for use in **geometry**. One of the angles of a set square is always a **right angle**.
The draughtsman checked all the corners with a set square.

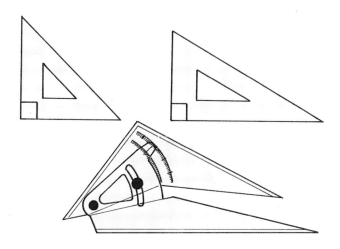

shape (two-dimensional) ► page 128

shape (three-dimensional) ► page 130

share *noun*
A share is a part of the **capital** of a company. In order to raise money, a company might sell some shares. The people who buy the shares, called **shareholders**, will then own part of the company. If the company is successful, those with shares will receive some of its **profits** and will be able to vote on what the company should do. Shares are bought and sold on a **stock exchange**.
They each paid 30 pence for each share.

share price *noun*
The share price is the price of a particular share.
The share price of the company is £1.35.

shareholder *noun*
A shareholder is someone who owns **shares** in a company. Another term for a shareholder is **member**.
The company paid a dividend to its shareholders last year.

shekel *noun*
The shekel is the currency of Israel.

shop *noun*
A shop is a place where customers are able to buy goods or services. It is usually a small business dealing with a limited number of goods. A large shop is often called a store.
He bought a television at the shop.

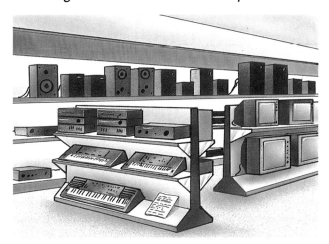

SI units *noun*
SI is short for Système International d'Unités, or International System of Units. It is a system of **measurement** used by many countries to allow them to compare scientific results. The most common SI units are the three bax units. The SI unit of length is the metre, the SI unit of mass is the kilogram and the SI unit of time is the second.
They used SI units to measure the imported cartons of tomatoes.

shape (two-dimensional) *noun*

A shape is any form. A two-dimensional shape is the outline of a flat surface. These shapes are also called **plane figures**. If they have straight sides they belong to the group of shapes called **polygons**.

Circles, squares and triangles are all two-dimensional shapes.

polygon

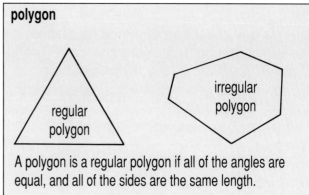

A polygon is a regular polygon if all of the angles are equal, and all of the sides are the same length.

triangles
Triangles are 3-sided figures.

quadrilaterals
Quadrilaterals are 4-sided figures.

parallelograms

Quadrilaterals with opposite sides parallel are parallelograms.

In theory, there is an infinite number of polygons, each with one more side than the last. Not all of these have been named.

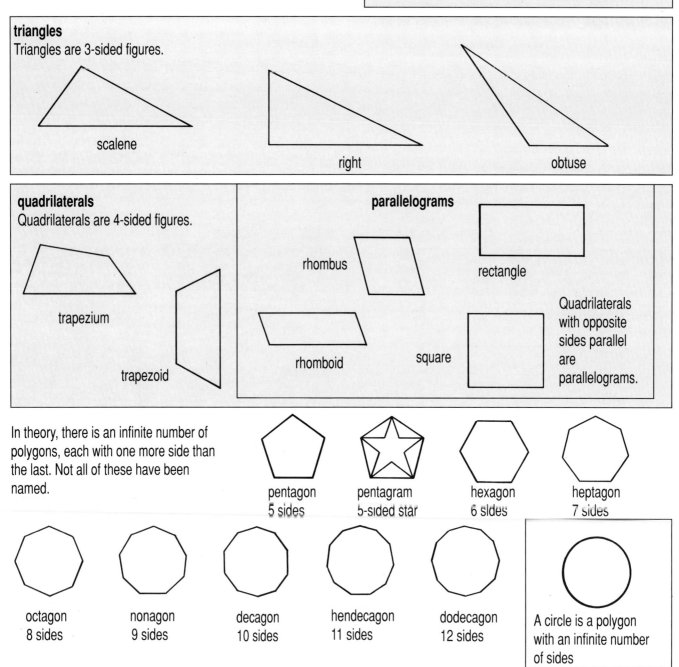

pentagon
5 sides

pentagram
5-sided star

hexagon
6 sides

heptagon
7 sides

octagon
8 sides

nonagon
9 sides

decagon
10 sides

hendecagon
11 sides

dodecagon
12 sides

A circle is a polygon with an infinite number of sides

curves

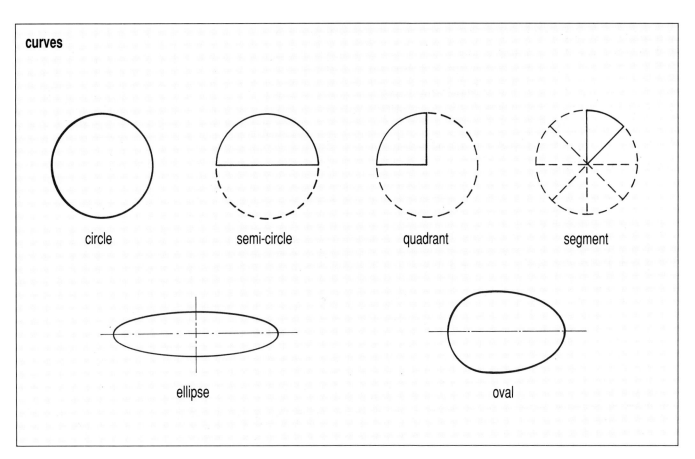

circle semi-circle quadrant segment

ellipse oval

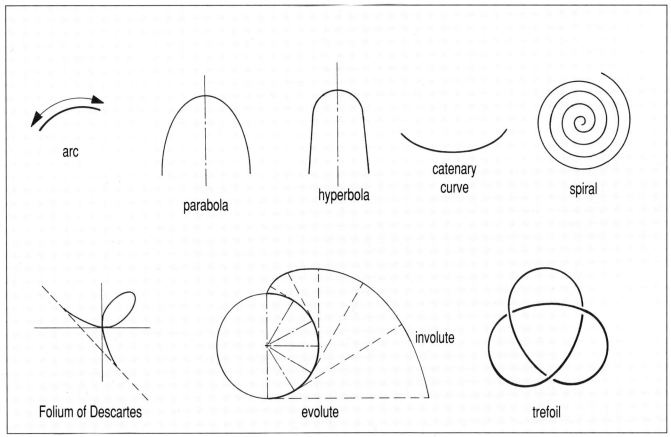

arc

parabola hyperbola catenary curve spiral

Folium of Descartes evolute involute trefoil

shape (three-dimensional) *noun*

A shape is any form. A three-dimensional shape is also known as a **solid**. Each side of a solid is called a face. If the faces are flat, the shape belongs to a group called **polyhedrons**.

Spheres, cuboids and pentahedrons are all three-dimensional shapes.

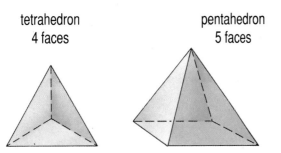

tetrahedron
4 faces

pentahedron
5 faces

polyhedron

regular polyhedron

irregular polyhedron

A polyhedron is a three-dimensional solid with flat faces. If all of the faces are exactly the same, it is a regular polyhedron. There are only five regular solids, tetrahedrons, cubes, octahedrons, dodecahedrons and icosahedrons.

---

hexahedrons
Hexahedrons have 6 faces.

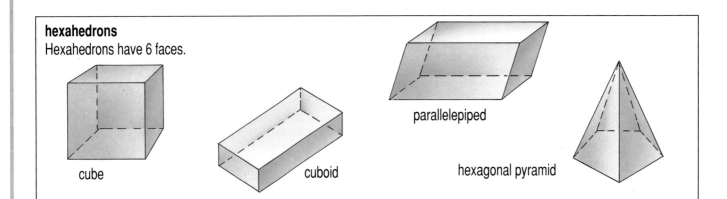

cube

cuboid

parallelepiped

hexagonal pyramid

---

heptahedron
7 faces

octahedron
8 faces

decahedron
10 faces

dodecahedron
12 faces

icosahedron
20 faces

---

prisms
Prisms have two parallel faces which are polygons. All the other faces are parallelograms.

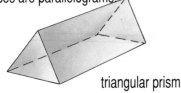 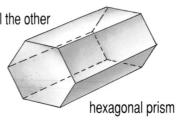 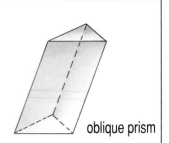

triangular prism

hexagonal prism

oblique prism

curved solids

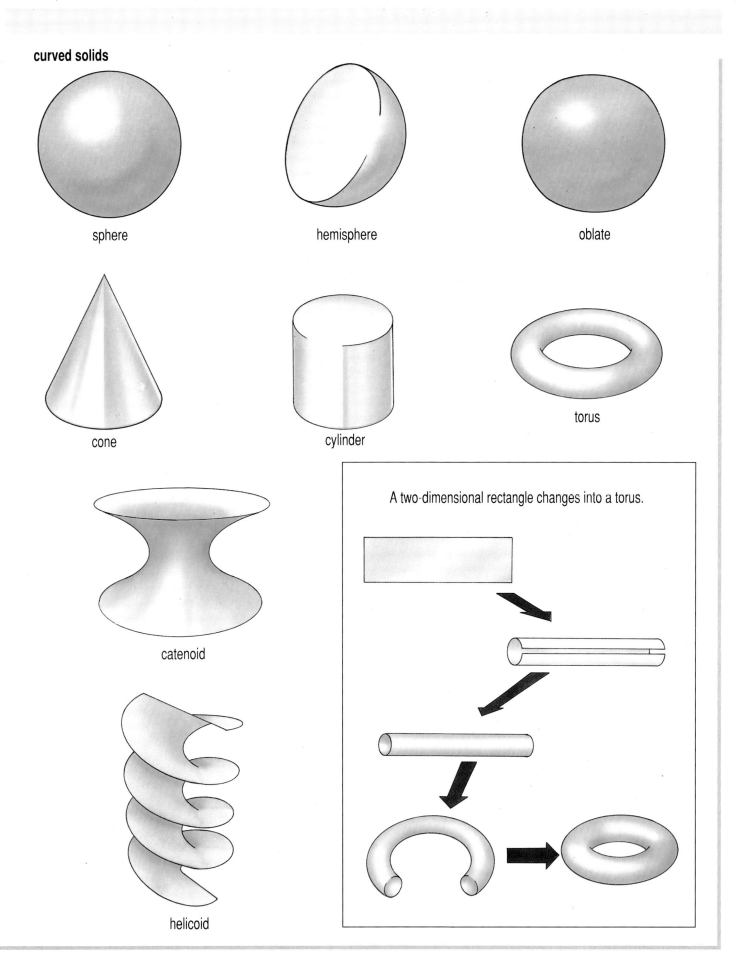

sphere

hemisphere

oblate

cone

cylinder

torus

catenoid

helicoid

A two-dimensional rectangle changes into a torus.

131

significant figures   *plural noun*

Significant figures are the **digits** that you keep after **rounding**. If 23.45886 is rounded to four significant figures, only the first four digits are kept, so the answer will be 23.46. The fourth **digit** is rounded from 5 to 6.
The value of pi is 3.1416 to five significant figures.

signs and symbols   ► page 133

silver   *noun*

Silver is a valuable metal. It is dug out of the ground from a silver mine. Silver was once used to make coins, but now most coins are made of other metals. Silver is used mostly in the manufacture of jewellery.
She bought a necklace made of silver.

similar   *adjective*

In **geometry**, similar describes things that are the same shape but different sizes.
If two triangles are similar, the larger one can be reduced to the size of the smaller one.

simple interest   *noun*

Simple interest is **interest** which is calculated on the amount of a original loan.
He paid simple interest on the loan, which kept costs down.

simplify   *verb*

To simplify means to write in a shorter form. A sequence of numbers in a problem can be simplified in order to solve it more easily. For example, 3 + (5 + 1) can be simplified 3 + 6.
It was necessary to simplify the number sequence to arrive at a single digit.

simultaneous equations   *noun*

Simultaneous equations are two or more equations containing more than one unknown quantity. They can be solved by drawing them on a graph, or by adding the equations together to eliminate the unknown one by one, using **substitution**.
They solved the problem using simultaneous equations and a graph.

sine   *noun*

The sine is the length of the side of a right-angled triangle which falls opposite an **acute** angle, divided by the length of the hypotenuse. It is one of the three main **functions** in **trigonometry**.
They measured the sides of the right-angled triangle to work out the sine.

Singapore dollar   *noun*

The Singapore dollar is the currency of Singapore. It is made up of 100 cents.

size   *noun*

Size is a word that describes how big something is.
A solid object needs three dimensions to describe its size.

slide rule   *noun*

A slide rule is a kind of **calculator** designed to work out mathematical problems. It is a ruler with a sliding section in the middle. **Scales** marked on each section must be lined up properly to provide the answer to a problem. The answer is read from one of the scales when it lines up with a line on the sliding section, or cursor.
Slide rules have now been replaced by electronic calculators which are faster and more accurate.

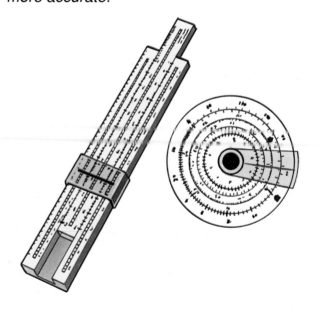

signs and symbols   *plural noun*

Signs and symbols are marks which stand for something else. They often take the place of words, and are quicker to write. Each sign or symbol has a special meaning, but sometimes they have more than one meaning. For example 12' can mean 12 feet or 12 minutes.

Signs and symbols are common in mathematical equations.

Business and commerce

$	dollar
¢	cent
£	pounds sterling
ª/c	account
@	at
©	copyright
®	registered trademark

Many mathematical signs symbols are also used in business and commerce.

Mathematics

+	plus or positive
−	minus or negative
x	multiplied by
•	multiplied by
÷	divided by
=	equal to
≈	approximately equal to
≡	equivalent to
≅	congruent or approximately equal to
≠	not equal to
<	less than
>	greater than
≤	less than or equal to
≥	greater than or equal to
∞	infinity
⊥	perpendicular to
||	parallel to
:	ratio, divided by
±	plus or minus
°	degree
'	minute
"	second
'	foot
"	inch
∴	therefore
π	pi
√	square root

The equal to sign is used in equations.

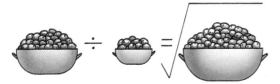

This equation uses division, equal to and square root signs.

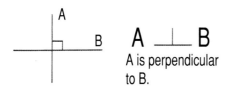

A is perpendicular to B.

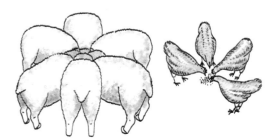

The ratio sign is used to compare two quantities. Six sheep to four chickens is written 6:4.

The ® symbol shows that a trademark has been registered.

133

slope *noun*

The slope of a surface describes the angle at which it drops away from a level surface. The slope of a hill is worked out by measuring the height of the angle compared with the distance travelled. If you rise one metre for every four metres you move, the slope is ¨, sometimes written as 1 in 4, or 25 per cent. The slope of a level surface is zero. Gradient is another word for slope.

The car could not drive up a slope of 40 per cent.

slump *noun*

A slump is a period of poor sales, or **recession**. Individual businesses can experience a slump, and so can whole countries.

He lost his job during the slump.

social security *noun*

Social security is a fund of money raised by a government and used to make payments to people who do not have enough money to live. The money is raised from compulsory contributions made by employees and employers. It can also be used to provide **pensions** and medical aid to those who are in need.

When they lost their job at the mine, they had to rely on social security.

solid *noun*

A solid is a **shape** which has **three dimensions**, length, width and height. Most solid shapes can be described as **polyhedra**.

A sphere is a solid, and a circle is a plane.

solution *noun*

A solution is the answer to a problem or question. For example, in the equation $3 = x + 1$, the solution is 2. It is written as $x = 2$.

The solution of the equation $4 = x + 7$ is $x = -3$.

solve *verb*

solvency *noun*

Solvency describes the ability of a person or a company to pay any debts. The opposite of solvency is **insolvency**.

Because of its solvency, the company was able to pay the huge shipping bill.

solvent *adjective*

souk *noun*

A souk is a kind of **market** often found in the Middle East and North Africa. It usually holds indoor stalls selling different sorts of food and goods.

The tourist bought a gold necklace at the souk in Fez.

spend *verb*

To spend is to pay out money in exchange for goods or services. The opposite of spend is save.

She spent more than she could afford on her new camera.

spendthrift *noun*

A spendthrift is someone who spends money carelessly and does not try to save it. The opposite of a spendthrift is a **miser**.

The woman was a spendthrift because she spent all her money as soon as she earned it.

sphere *noun*

A sphere is a perfectly round, **solid** shape. If a sphere is cut in two, the outline of the dissected base is a circle. A ball is an example of a sphere.

They built a communications satellite in the shape of a sphere.

spherical *adjective*

spiral *noun*

A spiral is a type of curve that winds round a fixed point, steadily moving away from it.
The paper streamer was spiral shaped.

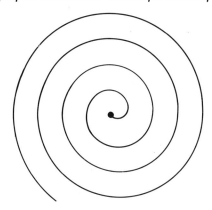

spreadsheet *noun*

A spreadsheet is a sheet of paper or computer-generated **table** of numbers in rows and columns. It is used by accountants to help **analyse** data more easily.
The company accountant used a spread sheet to compare the sales data on various products.

square *noun*

1. A square is a **rectangle** that has four equal sides. The angles of a square are **right angles**.
The lawn was shaped like a square.
2. The square of a number is the number multiplied by itself. The square of 3 is $3 \times 3 = 9$. The square of 3 is written 3^2.
The square of 5 is 25.
3. A square is a device used for measuring right angles. It is commonly known as a **set square**.

square measure *noun*

Square measures are measurements of area. The area of a football ground is worked out in square metres, or m^2. Square measures are compared to each other in size, eg: $100 \ cm^2 = 1 m^2$
$1,000 \ mm^2 = 1 m^2$.
The different fields were compared in size using square measures.

square root *noun*

A square root of a number is a number which when multiplied by itself gives the required number. The square root of 25 is 5, because $5 \times 5 = 25$. The symbol for square root it $\sqrt{}$. For example $\sqrt{25} = 5$.
He worked out the square root of 81 on his calculator.

standard *noun*

1. A standard is something that is used as the basis for comparison. It is used as a model by which other similar things are judged.
The deluxe pillow has more feathers than the standard pillow.
2. A standard is an accepted, often legal measurement used to compare the accuracy of other measurements. A standard is also a measure of value for many **commodities** such as gold. Because the value of gold is fixed, the value of other commodities can be measured against it.
The standard used by the country is the gold standard.

standardize *verb*

standard of living *noun*

The standard of living is a measure of what people can afford to buy with their money. In a wealthy country, where there is plenty of money to spend and people buy more, the standard of living is said to be high. In poor countries, where wages are low and people cannot spend much, the standard of living is said to be low.
The standard of living in the country rose as it became richer from its oil revenues.

standing order *noun*

A standing order is a request to a **bank** to pay a certain sum of money at regular intervals. A standing order may be used to pay off a **loan** over a number of months or years. Standing order payments are normally made each month.
Banks often take responsibility for paying standing orders on behalf of their clients.

star *noun*
A star is a flat shape that has sharp points. Stars can have any number of points, but usually they have five or six.
A pentagram is a star that has five points.

statement *noun*
A statement is a summary of an **account**. It is usually issued monthly or quarterly and tells how much a person or company owes, how much is owed and how much cash is in **reserve**.
The statement showed that the company owed the builders £100.

statistics *plural noun*
Statistics are collections of facts and **data** based on numbers. Statistics can be used to forecast how many people might buy a particular kind of product, or which government party might get the most votes during an election. Usually, the information is collected by researchers. Statistics is also used to mean the science of collecting and using the facts.
The statistics showed that 30 per cent of people prefer to travel by bus.

sterling *adjective*
Sterling describes the currency of the United Kingdom.
An English pound is a pound sterling.

stock *noun*
1. Stock is the **goods** bought or made by a business to sell to its customers, or to use in making other goods. The toys in a shop are its stock. A toy manufacturer buys a stock of plastic, wood and other materials to make the toys.
The stock of the department store was valued at £500,000.
2. Stock is a kind of **investment**. Usually, the word means money lent to a government. The government pays interest to the **investor** until the money is repaid.
She made a $5,000 investment in government stock.

3. In Britain, and more often in the United States of America, the word stock can refer to shares in a company.
Stocks and shares can be bought and sold on a stock exchange.

stock exchange *noun*
A stock exchange is a place where **stocks** and **shares** are bought and sold.
Stockbrokers buy and sell shares on the stock exchange for their customers.

stock market *noun*
A stock market can mean the same as **stock exchange**.
The stock market improved as several companies showed huge year end profits.

stockbroker *noun*
A stockbroker is a person whose job it is to buy and sell **stock** or **shares** on behalf of **clients**.
The stockbroker received a large commission for selling the shares at a high price.

stone ► **weights and measures**

store ► **shop**

straight angle *noun*
A straight angle is an angle that measures exactly 180°.
The two arms of the fan made a straight angle when the fan was fully opened.

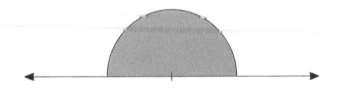

straight line *noun*
A straight line is a line that has no bends in it. Straight lines can be drawn with a ruler.
He drew a straight line 20 centimetres long.

strategy *noun*
A strategy is a plan. Companies often work out a strategy for selling their goods in a particular **market**.
The sales strategy the company used led to hugely increased sales in Africa.

strike *noun*
A strike is an agreement between employees to stop working. Usually they do this to put pressure on their employer to make some change. Employees may call a strike to get better working conditions or because they feel they are poorly paid.
The workers organized a strike for higher wages.
strike *verb*

subset *noun*
A subset is a **set** that is part of a larger set. In a set of pencils of different colours, the pencils can be grouped into subsets of the same colour.
The set of blue pencils is a subset of the set of all the pencils.

subsidiary *noun*
A subsidiary is a company that is controlled by another company.
Subsidiaries are sometimes created by a takeover.

subsidy *noun*
1. A subsidy is a sum of money paid to companies which need financial help to stay in business. For example, a government might pay a subsidy to the motor industry so that it can sell cars at a competitive price to **imported** cars.
Building companies have received a large government subsidy.
2. A subsidy is a sum of money paid to finance an organization which does not make a **profit**. A government might pay a subsidy to a hospital so it can provide medical services.
Many schools have received government subsidies in the past.

substituting *verb*
Substituting means replacing one thing by another in an **equation**. If 2 is substituted for x in the equation $3 = x + 1$, the equation becomes $3 = 2 + 1$. Both sides of this new equation are equal, so the solution of the equation is $x = 2$.
This equation was solved by substituting 2 for x.
substitution *noun*

subtract *verb*
To subtract means to take away. Its sign is $-$. $10 - 8 = 2$ is a subtraction problem.
He had to subtract the price of the cake from his pocket money.
subtraction *noun*

sum *noun*
1. The sum of two numbers is the total obtained when the numbers are added together.
The sum of 4 and 5 is 9.
2. A sum is a problem in arithmetic.
The class spent an hour practicing sums.

supermarket *noun*
A supermarket is a large store usually selling groceries and other household goods. Customers help themselves from the shelves. Goods are paid for at the **checkout**.
The shelves in the supermarket were piled high with groceries.

137

survey *verb*

1. To survey is to examine a subject by asking questions. Businesses often carry out a survey to find out who and where their customers are, and what they want to buy. *The company plans to survey book-buying habits in the city.*

2. To survey is to measure and map a certain area. A survey includes details of the boundary, elevation and position of the things in the area. **Geometry** and **trigonometry** are used in making the measurements. *A local company was hired to survey the building site.*

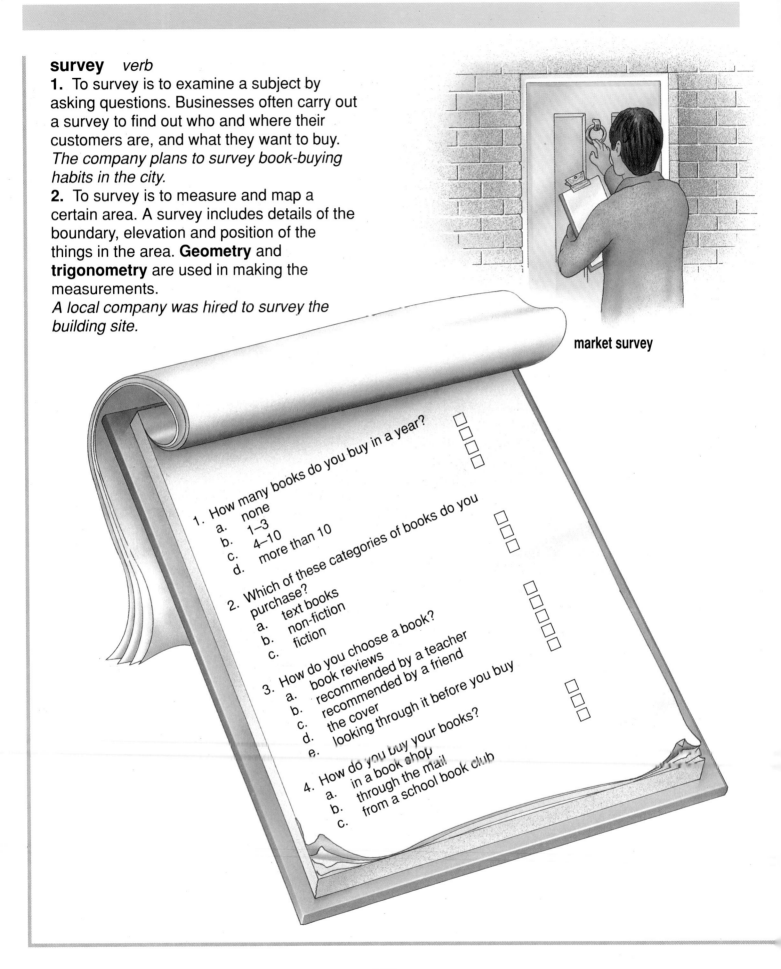

market survey

1. How many books do you buy in a year?
   a.   none
   b.   1–3
   c.   4–10
   d.   more than 10

2. Which of these categories of books do you purchase?
   a.   text books
   b.   non-fiction
   c.   fiction

3. How do you choose a book?
   a.   book reviews
   b.   recommended by a teacher
   c.   recommended by a friend
   d.   the cover
   e.   looking through it before you buy

4. How do you buy your books?
   a.   in a book shop
   b.   through the mail
   c.   from a school book club

land survey

tape survey

The distance between A and B can be measured by stretching a steel tape horizontally between the two points.

tacheometry
(stadia) survey

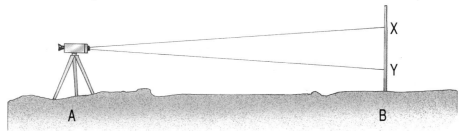

The distance between the instrument A and the staff B can be calculated by the difference between the two stadia readings, X and Y.

EDM survey

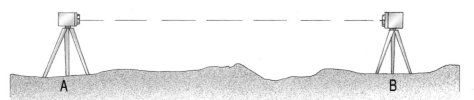

An EDM (electronic distance measurement) instrument at point A sends a signal to a target at point B. It then calculates the distance from the time it takes for the signal to travel.

levelling

Levelling is a method of measuring the difference in height between points. The surveyor looks through an instrument called a level at a measuring stick called a staff. A measurement is taken of the height showing on the staff at point A. Then the staff is moved to point B, the level swung around, and another measurement is taken. The surveyor can then calculate the height of the instrument and the height of point B.

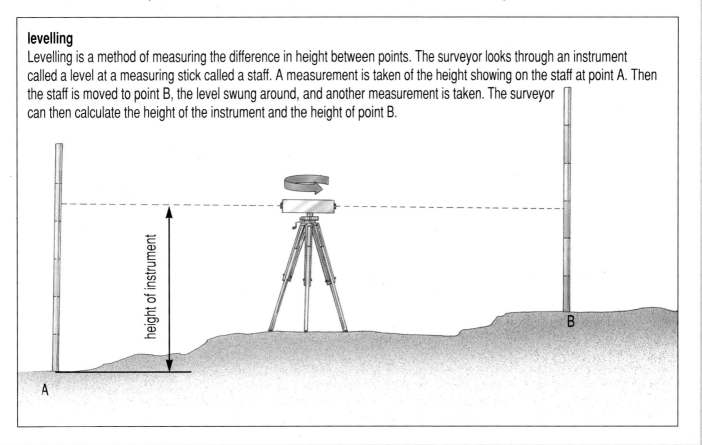

supplementary angles   *plural noun*
Supplementary angles are angles that add up to 180°.
A 60° angle and 120° angle are supplementary angles.

supplier   *noun*
A supplier is a person or a company that sells goods or services to customers.
The supplier delivered the fresh fruit to the supermarket each day.

supply   *verb*
To supply means to provide goods or services to customers.
The company said it would supply the supermarket with fresh fruit each day.

supply and demand   *noun*
Supply and demand is a system used in the market place. Customers freely decide which goods they do or do not wish to buy. If people want a particular item, then there is a demand for it. If a company can supply the item, then high sales will result. If there is poor demand or if a company cannot supply it, then sales will be low.
Because of supply and demand the company sold most of its stock.

surface   *noun*
A surface is the outer face of an object. The surface of a cube is made up of six square faces.
The Earth's surface, known as its crust, is between 8 and 40 kilometres thick.

surplus   *noun*
A surplus is the money left over from a total sum after payments for expenses have been made. Surplus money is often **profit**.
The charity made a surplus of £240 on the sale of flags.

survey ► page 138

Swiss franc   *noun*
The Swiss franc is the currency of Switzerland. A franc is made up of 100 centimes.

symbol   *noun*
A symbol is a **sign** that stands for something else. A symbol is often an abbreviation of the thing it stands for. For example, the symbol for degree is the sign ° and the symbol for metres is m. Symbols are used to represent unknown numbers in **algebra**.
We all recognize the symbol ÷.

symmetry ► page 141

syndicate   *noun*
A syndicate is a group of people or companies. It is often formed when several companies join together in order to buy something too expensive for the individual company to buy alone.
The companies formed a syndicate to build the new communications system.

symmetry *noun*

Symmetry means being the same on both sides. A rectangle has symmetry because a line drawn down the centre creates two identical sides. A circle has symmetry because both sides of the **diameter** match. The line which can be drawn to divide the object is the **axis** of symmetry.

A line of symmetry can be drawn down the centre of a butterfly's body.

symmetrical *adjective*

If a line was drawn through the middle of the Eiffel Tower from top to bottom, both sides of the line would show symmetry.

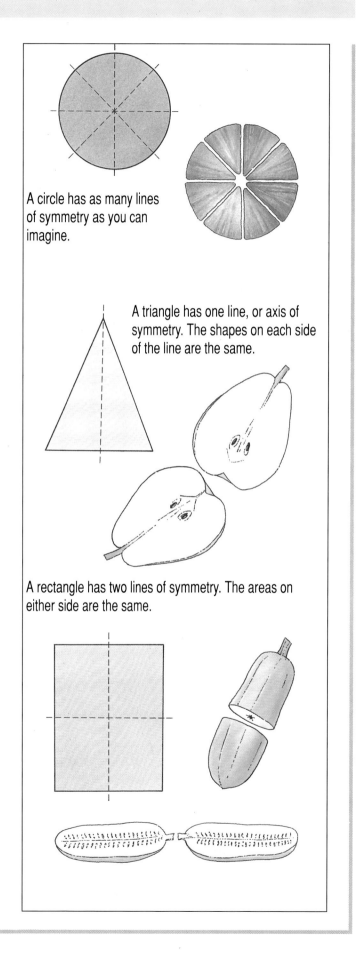

A circle has as many lines of symmetry as you can imagine.

A triangle has one line, or axis of symmetry. The shapes on each side of the line are the same.

A rectangle has two lines of symmetry. The areas on either side are the same.

141

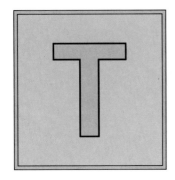

takeover *noun*
A takeover is when one company buys all or most of the **shares** of another company. The company that buys these shares will then control the other company.
A company that is making a loss may not be able to stop a takeover by another company.

talent *noun*
A talent was a unit of weight and of money used in ancient times.

tangent *noun*
1. A tangent is a straight line that touches a curve but does not cross it.
A rocket is launched into space at a tangent to the Earth's surface.
2. The tangent of an angle in a **right-angled triangle** is the **ratio** produced when the length of the side opposite the angle is divided by the length of the side **adjacent** to the angle.
The tangent of the triangle was 3:5.

tangram ► page 143

tariff *noun*
1. A tariff is a tax that is paid when goods are **exported** or **imported**. For example, a government might want people to buy cars that are made in their own country. It could place a tariff on imported cars so that they became more expensive than the country's own cars.
The government put a tariff on imported washing machines.
2. A tariff is a list of the prices charged by a business for the goods and services it sells.
The shopkeeper displayed a tariff on the wall of his shop.

Taxi service:	$
minimum charge	1.50
2–3 miles	2.80
4–5 miles	4.10
6–10 miles	8.00
over 10 miles	price on request

tax *noun*
Tax is money paid by people and companies to the government. The amount of **income tax** that people pay depends on how much income they earn. The amount of tax that a company pays depends on how much **profit** it earns. The government also raises money from **sales tax**. The government uses taxes to provide services such as roads, hospitals and schools.
The company paid 25,000 francs tax.

taxation *noun*
Taxation is the act by a government of charging taxes. It also means the money, or **revenue**, raised from taxes.
Without taxation, the government could not build roads.

taxpayer *noun*
A taxpayer is a person who pays tax.
The taxpayer asked his accountant to work out his tax bill.

tangram *noun*
A tangram is an ancient Chinese puzzle consisting of a square cut into seven pieces. These are five triangles, a square and a **parallelogram**. The seven pieces can be combined to form many different figures and shapes.
She made a bird with the tangram.

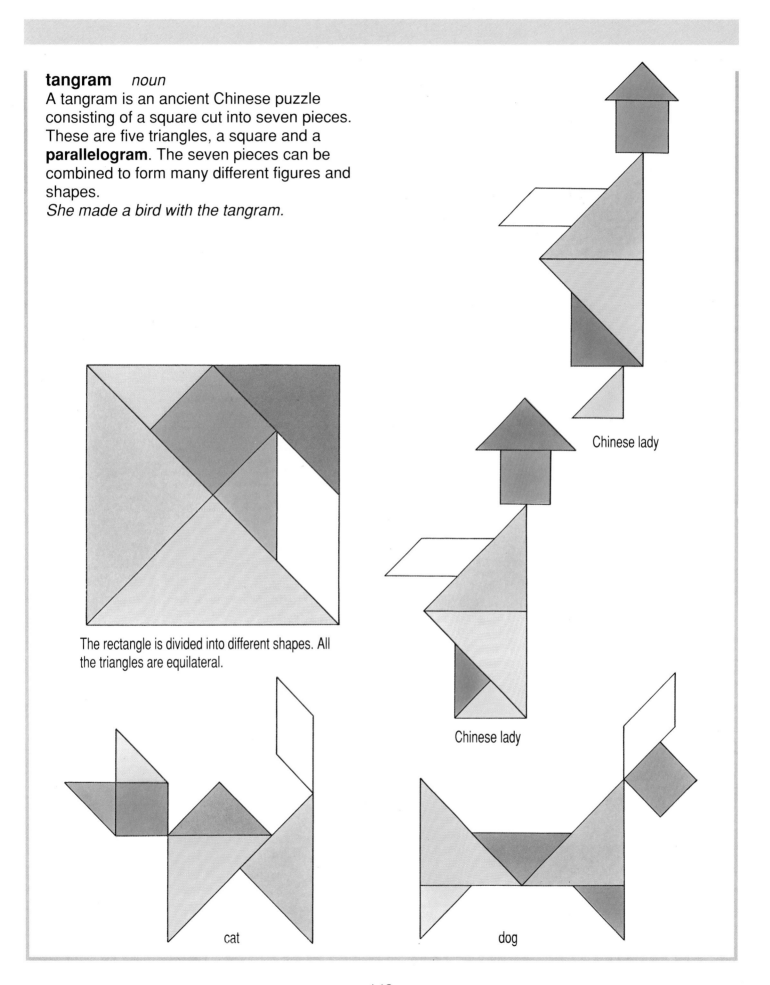

The rectangle is divided into different shapes. All the triangles are equilateral.

Chinese lady

Chinese lady

cat

dog

telecommunications   *plural noun*
Telecommunications are methods of communicating over large distances. This is made possible by means of modern information technology, such as communications satellites. Telephone calls, faxes and messages sent by computer are examples of telecommunications.
He used the telecommunications network to speak to his friend in New York.

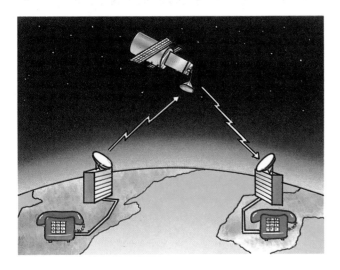

temperature   *noun*
Temperature is a **measurement** of how hot or cold something is. Temperature is normally measured using a **thermometer**. This records degrees of heat and cold on the **Celsius** or **Fahrenheit** scale.
The chef checked the temperature of the oven by looking at the thermometer.

tenant   *noun*
A tenant is a person or company who occupies a building that belongs to the owner, called the landlord. A tenant pays **rent** to the landlord, or has a **lease** agreement.
The tenant paid rent of £200 per month to the landlord.
tenancy   *noun*

tender   *verb*
To tender means to offer money to pay for something.
He tendered his bus fare.

tender   *noun*
A tender is an offer to supply goods or services. The company making the tender will explain the price it intends to charge, and the date when the work will be completed. If a company needs a new office block, it will ask several builders to say how much they will charge to build it. The builders will reply by putting in tenders.
The company employed the builder who put in the lowest tender.

tessellation ► page 146

tetrahedron   *noun*
A tetrahedron is a solid shape that has four faces. Each face is a triangle.
A tetrahedron is a polyhedron.

tetromino   *noun*
A tetromino is a flat, or one-dimensional, **shape** made by joining four squares together.
A tetromino is a polyomino that can be used in tessellation.

theft   *noun*
Theft means taking something without the owner's permission. Theft is a crime.
The man was put in prison for five years for theft.

three-dimensional   *adjective*
Three-dimensional describes objects that are **solids**. Three **co-ordinates** are needed to locate a particular point on a three-dimensional object. These refer to height, length and width.
A brick is a three-dimensional object.

time   *noun*
Time is a **measurement** of how long something lasts, or the period between events. A clock is used to measure time in seconds, minutes or hours.
The match will begin in 45 minutes.

time zone ► page 148

timetable *noun*

A timetable is a **table** of figures that shows when events are due to happen. A railway timetable shows when each train is due to leave or arrive at a station.
A school timetable shows when each lesson is due to begin and what the subject will be.

tithe *noun*

A tithe was a kind of **tax** used in the past. It was paid by people who worked on the land. They had to pay a tenth of their produce or income to the landowner.
The peasants paid their tithes every year after the harvest.

token *noun*

A token is an object that stands for something else. For example, a token can be a small disc that is used instead of money. People can buy tokens to use in an automatic car wash.
He bought 10 tokens to pay for rides at the fun fair.

ton *noun*

A ton is an **imperial measure** of weight. One long ton is equal to 2,240 pounds. It is equivalent to approximately 1,016 kilograms. One short ton is equal to 2,000 pounds, or approximately 907 kilograms.
People used to buy coal by the ton.

tonne *noun*

A tonne is a **metric measure** of weight.
One tonne is equal to 1,000 kilograms.

trade *verb*

To trade is to buy and sell. Trade can also mean exchanging something that you have for something of equal value that belongs to another person. This kind of trade is known as **barter**.
He traded a video for a compact disc.

trade *noun*

A person's trade is the job that they do.
His trade was bricklaying.

trademark *noun*

A trademark is a symbol, a word or a phrase. A trademark is sometimes referred to as a **logo** and is used by a company to mark its products. Customers will recognize the source of a product by its logo.
The trademark of the airline was a bird in flight.

trader *noun*

A trader is a person or company buying and selling goods. Other words for trader are dealer and merchant.
The trader sold eight pairs of shoes.

transaction *noun*

A transaction is an activity carried out by a business. For example, selling a product to a customer, paying wages to an employee, and depositing money in a bank account are all transactions.
The transactions of a business are recorded in its books of account.

translation *noun*

A translation is a sideways movement of an entire object or figure. If an arrow is translated, the length and direction of the arrow stay the same, but the ends of the arrow move at different places.
A translation is used to make a pattern.

trapezium *noun*

A trapezium is a **quadrilateral** in which two of the sides are **parallel** and the other two are not parallel.
The lawn was shaped like a trapezium.

tessellation   *noun*

Tessellation is the covering of a flat surface with shapes that fit together exactly.
Equilateral triangles and **regular hexagons** can cover a surface without leaving any gaps.
John made a tessellation out of dominos.

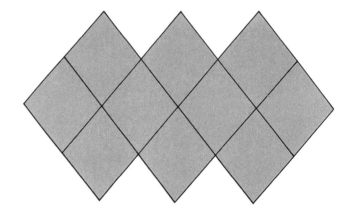

A tessellation using diamonds

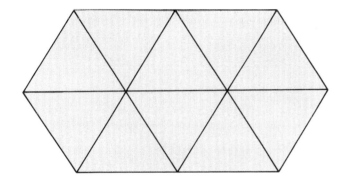

A tessellation using triangles

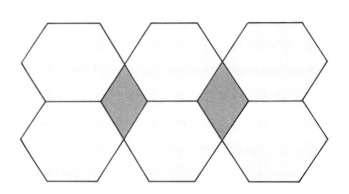

A tessellation using diamonds and hexagons

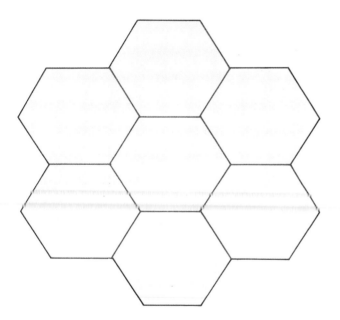

A tessellation using hexagons

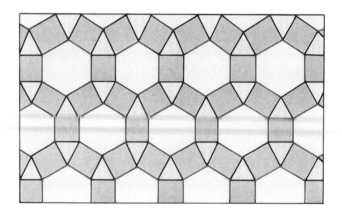

A tessellation using triangles, squares and hexagons

Tessellations using a variety of shapes

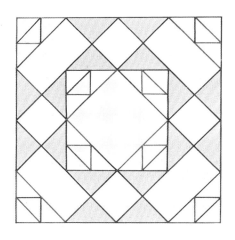

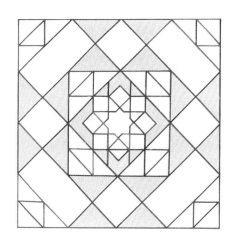

time zone *noun*

A time zone is a region of the Earth's surface where all the clocks tell the same time. When it is daytime on one side of the Earth, it is night on the other. As travellers move east or west, they pass through different time zones.

There are 24 time zones and they are counted east or west of Greenwich, in London, UK.

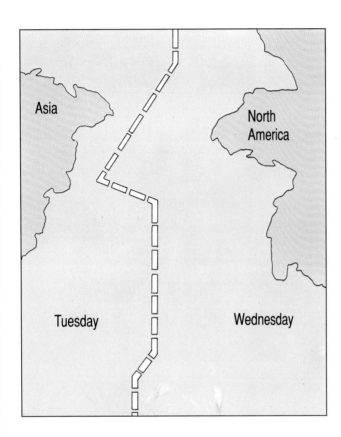

International date line

This imaginary line runs from north to south through the Pacific Ocean. For most of its distance it is exactly halfway round the world from Greenwich. It is where the day's date changes, and each new day begins. To the left of the line, it is a day earlier than it is to the right of the line.

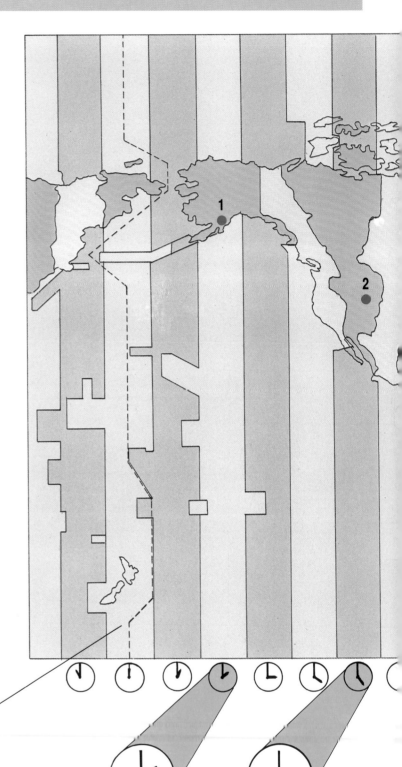

1 In Anchorage, Alaska, USA, it is 2.00 am. Everyone is asleep.

2 In Denver, Colorado, USA, it is 5.00 am. Early risers are starting to get up.

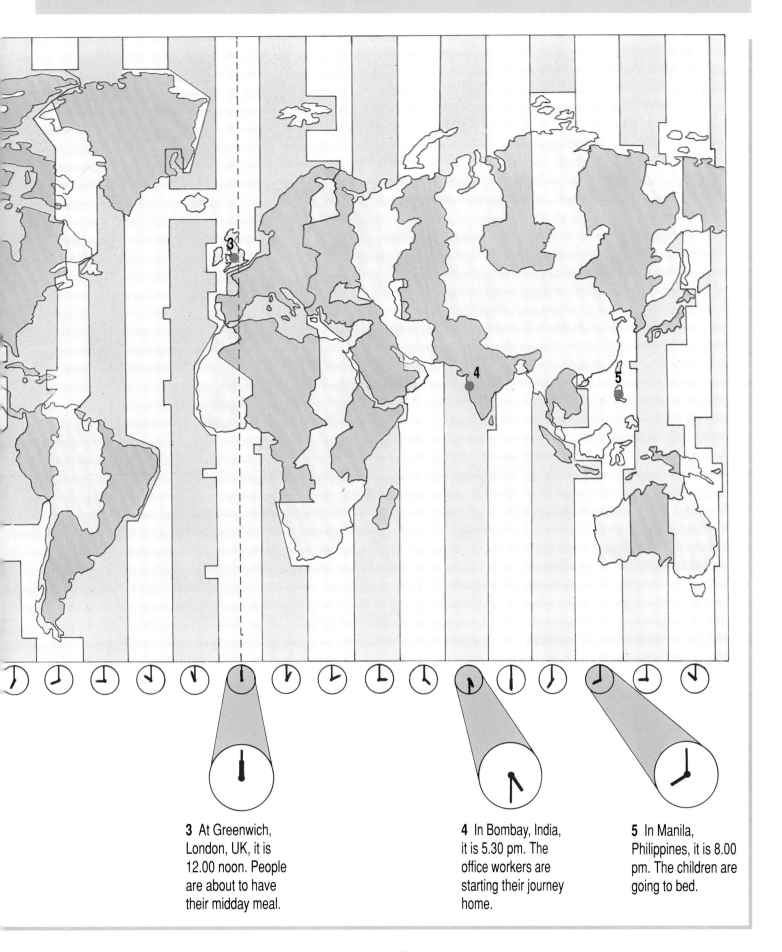

3 At Greenwich, London, UK, it is 12.00 noon. People are about to have their midday meal.

4 In Bombay, India, it is 5.30 pm. The office workers are starting their journey home.

5 In Manila, Philippines, it is 8.00 pm. The children are going to bed.

travellers' cheque *noun*
A travellers' cheque is a kind of cheque used by people who want to spend money in another country where the **currency** is different. To use a travellers' cheque it must be signed and then given to the foreign bank.
He cashed a travellers' cheque worth 100 francs.

treasurer *noun*
A treasurer is a person who is in charge of the money belonging to a business or organization.
The club's treasurer told the club members how much money the club had.

treasury *noun*
1. A treasury is a place where money and valuable objects are kept.
He added £50 to the club's treasury.
2. The treasury is the name sometimes given to the department in a government that is in charge of the country's money.
Taxes are paid to the treasury.

trend *noun*
A trend is the direction in which something is moving. Companies often work out business trends so they can predict future sales.
The fashion trend is towards red hats.

triangle *noun*
A triangle is a flat, or plane, shape that has three sides. There are four main types of triangle. These are a **right-angled triangle**, an **equilateral triangle**, an **isosceles triangle**, and a **scalene triangle**.
The interior angles of a triangle always add up to 180°.

triangular numbers *noun*
A triangular number is a number which can be drawn as a triangle of dots. One is the first triangular number and three and six are the next.
The first five triangular numbers are 1, 3, 6, 10 and 15.

triangulation *noun*
Triangulation is the method of dividing an area into triangles in order to **survey** it. **Trigonometry** is used in trinagulation.
The surveyor used triangulation to measure the site of the new house.

trigonometry *noun*
Trigonometry is the study of the angles and sides of **triangles** and other figures. It uses **ratios** called the **sine**, **cosine** and **tangent**.
He used trigonometry to calculate the angle of elevation.

trillion *noun*
A trillion is equal to one million million in many countries. It is written as a one followed by 12 zeros. In some countries, the original meaning of one million million million is still used. It is written as a one followed by 18 zeros.
There were more than a trillion grains of sand in the box.

1,000,000,000,000

trisect *verb*
To trisect is to divide into three parts. Each part is usually equal in size.
He used a pair of compasses to trisect the line.

tromino *noun*
A tromino is a shape made by joining three squares together along the edges.
A tromino is a polyomino that can be used in tessellation.

twenty-four hour clock   *noun*

The twenty-four hour clock counts the time straight through from 00.00 hours, or midnight, to 24.00 hours, which is also midnight. Most timetables use twenty-four hour time. Twenty-four hour times always have four figures. The first two show the hour, the last two show the minutes. For example, 04.15 shows that it is 4.15 in the morning, but 16.15 shows that it is 4.15 in the afternoon.

Airline departures follow a twenty-four hour clock timetable.

The twenty-four hour clock at Greenwich is unusual. Its hour hand goes around only once every 24 hours.

This digital clock uses a liquid crystal display face. It uses digits up to 24.00.

This clock face has two rings of numbers. It is easy to tell the time using either the 12 or 24 hour system.

An airport terminal clock shows when planes leave and when they arrive. They use a twenty-four hour clock so it is easy to tell if the time given is in the morning or afternoon.

trust fund   *noun*
A trust fund is an amount of money,
valuables or property that is set aside for
somebody's benefit. It may be set up by
parents for their children, or by insurance
companies for their clients.
*The trust fund made sure that the children
would have money to invest when they
reached the age of 21.*

turnover   *noun*
The turnover of a business is the amount of
money earned by that business from selling
goods or **services** to its customers. Another
word for turnover is **sales**.
*The turnover of the business increased by
£1.5 million last year.*

twenty-four hour clock ► page 151

two-dimensional   *adjective*
Two-dimensional describes objects that
are flat, such as a piece of paper. Two
co-ordinates are needed to describe a point
on a two-dimensional surface. These refer to
length and width.
A square is a two-dimensional shape.

underwrite   *verb*
1. To underwrite means to agree to accept a
financial **risk**. Underwriting is sometimes
used in connection with **insurance
companies** which agree to pay money in
the event of a disaster.
*The insurance company will underwrite his
policy of insurance against fire.*
2. To underwrite usually means to share the
risk when a company tries to sell its **shares**
to the public. If the public will not buy all the
shares at the required price, the underwriting
company promises to buy them instead.
*A group of insurance companies came
together to underwrite the issue of shares.*
underwriter   *noun*

unemployed   *adjective*
Unemployed is a word that describes people
who do not have a job.
*The number of unemployed rose when the
factory closed.*
unemployment   *noun*

union   *noun*
The union of two or more **sets** is the
collection of all the things in each set. If a
basket of fruit contains a set of apples and a
set of oranges, the union of the two sets
contains all the apples and all the oranges.
*The union of two sets can be worked out on
a Venn diagram.*

unilateral   *adjective*
Unilateral describes something that has only
one side.
*A straight line is an example of a unilateral
surface.*

unit *noun*

1. A unit is another word for one.
In the decimal system of counting, the numbers 1 to 9 are units because they are only one digit long.

2. A unit is a set or **standard**, or set quantity of **measurement**. Measuring something is the same as counting the number of units it has. **Metres** are units of length, and **degrees** are units of angle or temperature.
A gram is a unit of weight.

unit trust *noun*

A unit trust is a **fund**, usually managed by an **insurance company**, which pays money to buy **shares** and other **investments**. An individual may buy units in a unit trust and will receive an income from this investment.
He paid £800 to buy units in a unit trust.

United States dollar *noun*

The United States dollar is the currency of the United States of America. There are 100 cents in a dollar.

universal set *noun*

A universal set is a set made up of all the things that need to be included in a particular group because they share one or more **properties**. In a library, the universal set is the set of all the books on the shelves. These books can be grouped together into different topics, so that each topic forms a **subset** of the universal set of all books.
The universal set of chess pieces includes pawns, rooks, knights, bishops, queens and kings.

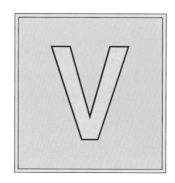

value *noun*

1. The value of a single number, or **digit**, is the amount it represents.
Six is greater in value than two.

2. The value of something is its price, or what it is worth.
The value of a diamond increases with size.

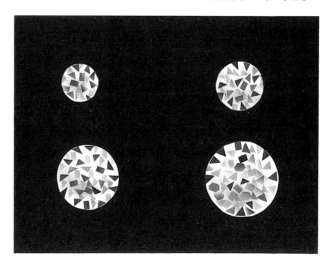

variable *adjective*

Variable is a word that describes things that change. The temperature of the sea is variable because it is warmer in summer than in winter. The opposite of variable is **constant**.
The speed of the bus was variable because it kept stopping and starting.

VAT ► sales tax

vault *noun*

A vault is a kind of large **safe**. It is used to store quantities of money or valuables. Vaults are usually made of steel.
The gold was stored in the bank's vault.

Venn diagram   *noun*

A Venn diagram is a way of drawing **sets** on a piece of paper. Each set is drawn inside a circle or other shape. If the contents of the sets overlap, those things in the overlapping section or **interset** can be identified as belonging to both of the sets.

Venn diagrams clearly show the things that are common to two sets.

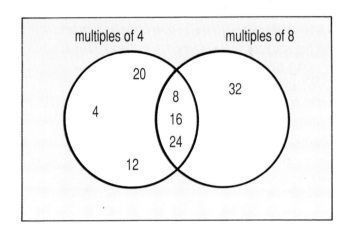

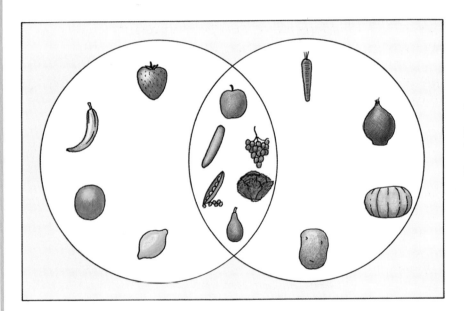

The contents of two sets, one of fruits and one of vegetables, overlap at the interset. The interset contains things common to both sets, in this case things which are green.

The contents of these three sets give four intersets.

1. The orange interset is common to the yellow and red sets.
2. The green interset is common to the yellow and blue sets.
3. The mauve interset is common to the red and blue sets.
The dark interset in the middle is common to all 3 sets.

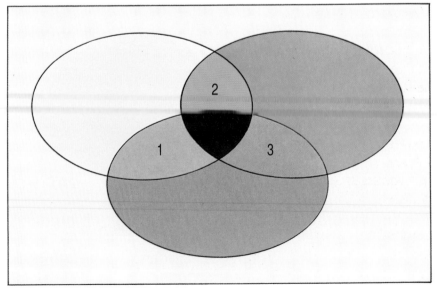

vector   *noun*
A vector is a quantity that has been worked out by knowing the size and direction of a moving object. The motion of an aircraft is an example of a vector. The speed of the aircraft is the size of the vector. The direction in which the aircraft is travelling is the direction of the vector.
They guided the pilot back to base once they had worked out the plane's vector.

velocity   *noun*
Velocity is another word for **speed**. It is a measurement of distance over time. Velocity can be measured in kilometres per hour. This is written in short as km/h, or as metres per second, m/s. In **Imperial measures**, it is written as miles per hour, or mph. Velocity is a **vector**.
In order to find the velocity of an object, you divide the distance the object has travelled by the time it took to do so.

Venn diagram ► page 154

venture capital   *noun*
Venture capital is money that is borrowed to start up a new **business**. This money is used for businesses that are expected to make a large profit, although there may still be a high **risk**.
Venture capital of 200,000 dollars was borrowed from the bank.

vertex (plural **vertices**)   *noun*
The vertex is the point where two lines meet. The corner point of a **polygon** or **polyhedron** is called a vertex. A vertex is also the point furthest away from the base of a shape. Another word for vertex is **apex**.
The vertex of a mountain is also known as its peak.

vertical   *adjective*
Vertical is a word that describes things that point straight up or down.
A flag pole stands vertical to the ground.

volume   *noun*
Volume is a **measurement**. It describes the amount of space an object takes up. It is measured in cubic units. The main unit of volume is the **cubic metre**, or m^3 for short.
Volume is the same as capacity and can also be measured in litres.

voucher   *noun*
1. A voucher is a document, or some other written evidence, that a payment has been made.
The voucher was evidence enough that the accounts were accurate and true.
2. A voucher is a document which can be exchanged for goods or services.
Many stores sell gift vouchers which can be used instead of money in any of their branches.

vulgar fraction ► **fraction**

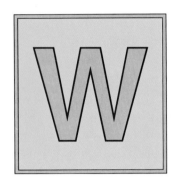

wages   *plural noun*
Wages are the money that an employee is paid to do a job. Wages means the same as **salary**.
His wages for last week amounted to 2,500 francs.

Wall Street   *noun*
Wall Street is another name for the New York Stock Exchange. The stock exchange is situated in Wall Street, in the Manhattan area of New York.
The value of shares traded on Wall Street yesterday was unusually high.

wampum   *noun*
Wampum is a collection of beads formerly used in **barter** by North American Indians. The beads were made from shells and were used as money and for jewellery. Black shells were worth more than white shells.
She bought the blanket with wampum.

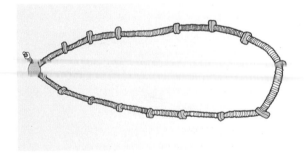

wealth   *noun*
Wealth is the accumulation of a quantity of something. It is usually used to mean the money and possessions owned by someone.
Most of his wealth came from his chain of restaurants.

week   *noun*
A week is a measurement of **time** which equals seven **days**. Many people work for five days of the week and have one or two days of leisure time. In many countries this rest period is known as the weekend.
It took four weeks to repair the school roof.

weight   *noun*
Weight is a kind of measurement. It measures the force with which an object is pulled towards the centre of the Earth. Weight is used to describe how heavy things are. The main unit of weight is the **gram**.
The weight of the can of beans was 450 grams.

weights and measures ▶ page 158

whole number ▶ **integer**

wholesale   *adjective*
Wholesale describes goods that are sold in large quantities to people or businesses who then sell them to the public. When a shop buys books from a publisher, it often buys a large number at once. The shop pays a lower price at the wholesale rate because it must then add on its costs and profit to resell at the higher **retail** price.
He bought his stock of winter shoes from the wholesale store.

will   *noun*
A will is a statement of how a person's property should be distributed after their death.
The will said that his house was to be given to his children.

withdrawal   *noun*
A withdrawal is money taken out of a bank account. Usually, only the owner of the account can withdraw money. The opposite of withdrawal is **deposit**.
He made a withdrawal of 25 pounds.
withdraw   *verb*

World Bank *noun*
The World Bank is an international bank that lends money to countries for development. It was founded in 1944 to help countries that had been economically ruined by the events of the Second World War.
The World Bank loan helped the country to build essential roads and bridges.

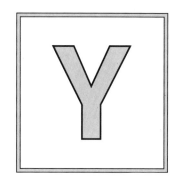

yard *noun*
A yard is an **imperial** measure of length or distance. There are 3 feet or 36 inches in a yard. It is also equivalent to the metric length of 90 centimetres.
A yard is approximately the length of a long stride.

year *noun*
A year is the time it takes for the Earth to travel once around the sun. This is $365\frac{1}{4}$ days. Every four years, there is a **leap year** which has an extra day, February 29th.
The Olympic Games are held every four years.

yen *noun*
The yen is the currency of Japan.

yield *noun*
The yield is the amount of money that is paid out on an **investment**. It is sometimes called the return earned on an investment.
The yield on their investment was more than they expected.

yuan *noun*
The yuan is the currency of China. The yuan is divided into 10 jiao or 100 fen.

weights and measures *plural noun*

Weights and measures are **measurements** used to show the size and weight of things. The most commonly used system of measurement in the world is the **metric system**, or SI system. The Customary system and Imperial system are still found, though the metric system has been officially adopted in most countries.

Some of the earliest weights and measures were calculated according to parts of the body, such as the foot or hand.

	imperial	American
1 pint	0.568 l	0.473 l
1 quart	1.137 l	0.946 l
1 gallon	4.546 l	3.785 l
1 peck (dry)	9.092 l	8.827 l
1 bushel (dry)	36.369 l	35.309 l

The imperial and American systems sometimes use the same terms to mean different amounts.

length and distance

imperial		metric	
1 inch (in)		1 nanometre (nm)	
1 foot (ft)	12 in	1 micron (μ)	1,000 nm
1 yard (yd)	3 ft	1 millimetre (mm)	1,000 μ
1 rod (rd)	$5\frac{1}{2}$ yd	1 centimetre (cm)	10 mm
1 furlong (fur)	40 rd, or $\frac{1}{8}$ mi	1 decimetre (dm)	10 cm
		1 metre (m)	10 dm
1 statute mile (mi)	5,280 ft	1 decametre (dam)	10 m
1 league (statute)	3 mi	1 hectometre (hm)	10 dam
		1 kilometre (km)	10 hm

surface and area

imperial		metric	
1 square inch (sq in)		1 square millimetre (mm^2)	
1 square foot (sq ft)	144 sq in	1 square centimetre (cm^2)	100 mm^2
1 square yard (sq yd)	9 sq ft	1 square decimetre (dm^2)	100 cm^2
1 square rod (sq rd)	$30\frac{1}{4}$ sq yd	1 square metre (m^2)	100 dm^2
1 acre (A)	160 sq rd	1 square decametre (dam^2)	100 m^2
1 square mile (sq mi)	640 A	1 square hectometre (hm^2)	100 dam^2
		1 square kilometre (km^2)	100 km^2

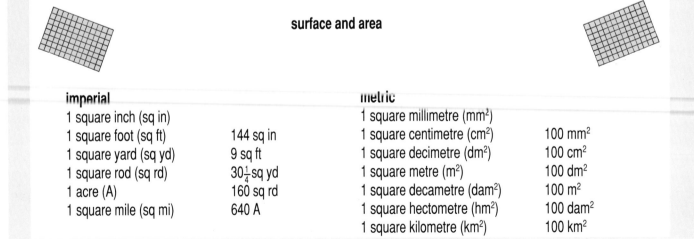

volume and capacity

imperial		metric	
1 cubic inch (cu in)		1 cubic millimetre (mm³)	
1 cubic foot (cu ft)	1,728 cu in	1 cubic centimetre (cm³)	1,000 mm³
1 cubic yard (cu yd)	27 cu ft	1 cubic decimetre (dm³)	1,000 cm³
		1 cubic metre (m³)	1,000 dm³
		1 cubic decametre (dam³)	1,000 m³
		1 cubic hectometre (hm³)	1,000 dam³

liquid measure

imperial		metric	
1 fluid dram (fl dr)		1 millilitre (ml)	
1 fluid ounce (fl oz)	8 fl dr	1 centilitre (cl)	10 ml
1 gill (gi)	5 fl oz	1 decilitre (dl)	10 cl
1 pint (pt)	4 gi	1 litre (l)	10 dl
1 quart (qrt)	2 pt	1 decalitre (dal)	10 l
1 gallon (gal)	4 qrt	1 hectolitre (hl)	10 dal
		1 kilolitre (kl)	10 hl

weight and mass

avoirdupois		metric	
1 grain (gr)		1 milligram (mg)	
1 dram (dr)	27.344 grains	1 centigram (cg)	10 mg
1 ounce (oz)	16 drams	1 decigram (dg)	10 cg
1 pound (lb)	16 ounces	1 gram (g)	10 dg
1 stone	14 pounds	1 decagram (dag)	10 g
1 hundredweight (cwt)	100 lbs (American)	1 hectogram (hg)	10 dag
1 hundredweight (cwt)	112 lbs (British)	1 kilogram (kg)	10 hg
1 short ton	2,000 lbs	1 quintal (q)	100 kg
1 long ton	2,240 lbs	1 metric tonne (t)	1,000 kg

zero   *noun*
Zero is the answer obtained when a number
is subtracted from itself. The sign for zero is
0. For example, 7 − 7 = 0.
Zero is the only number between the
positive numbers and the negative numbers.

Zimbabwe dollar   *noun*
The Zimbabwe dollar is the currency of
Zimbabwe. A Zimbabwe dollar is divided into
100 cents.

zloty   *noun*
The zloty is the currency of Poland. There
are 100 groszy in a zloty.